T0155513

SpringerBriefs in Electrical and Computer Engineering

Series editors

Woon-Seng Gan, Nanyang Technological University, Singapore, Singapore
C.-C. Jay Kuo, University of Southern California, Los Angeles, CA, USA
Thomas Fang Zheng, Tsinghua University, Beijing, China
Mauro Barni, University of Siena, Siena, Italy

SpringerBriefs present concise summaries of cutting-edge research and practical applications across a wide spectrum of fields. Featuring compact volumes of 50 to 125 pages, the series covers a range of content from professional to academic. Typical topics might include: timely report of state-of-the art analytical techniques, a bridge between new research results, as published in journal articles, and a contextual literature review, a snapshot of a hot or emerging topic, an in-depth case study or clinical example and a presentation of core concepts that students must understand in order to make independent contributions.

More information about this series at http://www.springer.com/series/10059

Guan Gui • Bin Lyu

Optimization for Wireless Powered Communication Networks

 Springer

Guan Gui
College of Telecommunications
and Information Engineering
Nanjing University of Posts
and Telecommunications
Nanjing, China

Bin Lyu
College of Telecommunications
and Information Engineering
Nanjing University of Posts
and Telecommunications
Nanjing, China

ISSN 2191-8112   ISSN 2191-8120   (electronic)
SpringerBriefs in Electrical and Computer Engineering
ISBN 978-3-030-01020-1   ISBN 978-3-030-01021-8   (eBook)
https://doi.org/10.1007/978-3-030-01021-8

Library of Congress Control Number: 2018957112

This Springer imprint is published by the registered company Springer Nature Switzerland AG
The registered company address is: Gewerbestrasse 11, 6330 Cham, Switzerland

Preface

Recently, wireless devices are deployed everywhere due to the development of Internet of Things (IoT). Wireless devices are typically powered by their embedded energy sources, e.g., batteries and supercapacitors. However, there usually exists an energy limitation problem for the long-term operation of wireless networks due to the limited capacities of these energy sources. Although the lifetime of wireless networks can be extended by recharing or replacing the energy sources, but it is usually inconvenient. Radio frequency (RF)-based wireless power transfer (WPT) is a promising technique to remotely power wireless devices, which can solve the energy limitation problem and prolong the lifetime of wireless networks. The integration of wireless communication networks and RF-based WPT has led to a novel networking paradigm, i.e., wireless powered communication network (WPCN), where wireless devices usually harvest energy from energy signals radiated by remote energy source (such as power beacons). In the past several years, WPCN has been one of the most hot research topics in the field of wireless communications. However, these are still some challenges for WPCN to be practically applied in IoT, such as low energy transmission efficiency and system throughput. The objective of this book is to present systematical mechanisms to realize system throughput maximization for WPCN.

In Chap. 1, we first give an overview of WPCN, namely, research background and recent advances. Then, we present two main working modes for WPCN, including the harvest-then-transmit (HTT) mode and the backscatter communication (BackCom) mode.

In Chap. 2, we investigate non-orthogonal multiple access (NOMA) in WPCN, which consists of a power station, two users, and an information receiver (IR). We consider the practical successive interference cancellation (SIC) constraints. To maximize the system throughput under SIC constraints, the optimal time and energy allocation scheme is designed.

In Chap. 3, we propose two hybrid schemes for WPCN, namely the hybrid user scheme and the hybrid mode scheme, which integrate the advantages of the HTT mode and the BackCom mode. In the hybrid user scheme, there are users adopting the HTT mode and users adopting the BackCom mode. The optimal time allocation among these users in investigated. Specifically, in the optimal time allocation policy, only the BackCom user with the maximum backscatter rate can be scheduled. In the hybrid mode scheme, each user can work either in the HTT mode or the BackCom mode. For this scheme, the optimal working mode permutation is derived.

In Chap. 4, we propose a novel hybrid mode scheme for cognitive WPCN (WPCN), which includes a primary communication system and a secondary communication system. Different from the hybrid mode scheme proposed in Chap. 3, in this chapter, the cognitive user (CU) in the secondary communication system can work either in the HTT mode or the hybrid BackCom mode, namely, the ambient backscatter mode and the bistatic scatter (BS) mode. The CU can harvest energy from both the primary transmitter (PT) and the power beacon (PB) following the HTT mode, while the PT and the PB serve as the incident energy sources for the AB mode and the BS mode, respectively. To maximize the system throughput, we investigate the optimal time allocation among the three modes, from which the optimal combination of working modes is presented.

In Chap. 5, we present a relay cooperation scheme for backscatter communication system (BCS), where the relay aids the user's information transmission. We consider two cases that the relay is with an embedded energy source and the relay is without an embedded energy source. For the previous case, the relay only uses its own energy for information forwarding, while for the latter case, the relay needs to harvest energy from the carrier emitter first and then use the harvested energy to forward the user's information to the IR. To maximize the system throughput, the optimal time allocation schemes are studied for both cases.

Finally, Chap. 6 summarizes this book and discusses the future research directions.

Nanjing, China   Guan Gui
Nanjing, China   Bin Lyu
July 2018

Acknowledgements

This work was partly supported by the Project Funded by the Priority Academic Program Development of Jiangsu Higher Education Institutions, National Natural Science Foundation of China Grants (No. 61701258), Jiangsu Specially Appointed Professor Program (No. RK002STP16001), Program for Jiangsu Six Top Talent (No. XYDXX-010), Program for High-Level Entrepreneurial and Innovative Talents Introduction (No. CZ0010617002), Natural Science Foundation of Jiangsu Province Grant (No. BK20170906), Natural Science Foundation of Jiangsu Higher Education Institutions Grant (No. 17KJB510044), NUPTSF (No. XK0010915026) and "1311 Talent Plan" of Nanjing University of Posts and Telecommunications.

Contents

Acronyms

AB	Ambient backscatter
BackCom	Backscatter communication
BB	Bistatic backscatter
BCS	Backscatter communication system
CE	Carrier emitter
CR	Cognitive radio
CU	Cognitive user
CWPCN	Cognitive wireless powered communication network
EB	Energy beamforming
EH	Energy harvesting
ET	Energy transmitter
FD	Full duplex
HAP	Hybrid access point
HD	Half duplex
HTT	Harvest-then-transmit
IoT	Internet of Things
MIMO	Multiple Input Multiple Output
NOMA	Non-orthogonal multiple access
PB	Power beacon
RF	Radio frequency
SIC	Successive interference cancellation
SWIPT	Simultaneous wireless information and power transfer
TDMA	Time division multiple access
WIT	Wireless information transmission
WPC	Wireless powered communication
WPCN	Wireless powered communication network
WPT	Wireless powered transfer

Chapter 1
Introduction

Abstract With the step growth of Internet of Things (IoT), energy-constrained wireless devices are deployed throughout our lives. Wireless powered communication network (WPCN) is a promising networking paradigm where wireless devices can be remotely powered by radio frequency (RF) enabled wireless power transfer (WPT) technology. In this chapter, we first given an overview of WPCN, which includes the research background and the recent research results. Then, we briefly discuss the main operation modes of WPCN.

1.1 Overview of Wireless Powered Communication Networks

With the rapid development of wireless communication, wireless devices are becoming more and more popular in daily life. At present, wireless devices usually rely on built-in batteries to supply power. However, due to the limited battery capacity, the lifetime of these devices is limited. For the devices such as mobile phones, these problems can be solved by replacing batteries or charging batteries, but these periodically operations can significantly reduce the user experience. While for low-power devices such as sensors deployed in Internet of Things (IoT), since the number of deployed sensors is huge and the sensors may be deployed in some dangerous scenarios, it may be inconvenient to replace batteries manually [1, 2]. Therefore, how to achieve wireless charging of wireless devices has become an urgent problem.

In order to solve the problem of wireless charging, energy harvesting (EH) technology has emerged [3, 4]. EH is a process in which wireless devices harvest energy from external energy sources and convert it into electrical energy. The energy sources available for EH include natural energy, such as solar [5], wind [6], thermal energy [7], and man-made energy sources, such as radio and television signals, WiFi signals [8]. Compared with the conventional energy supply methods, EH can effectively solve the problem of energy shortage of wireless devices. For example, in wireless sensor networks, EH technology is regarded as a good substitute for traditional batteries. The energy harvested by EH can effectively power the sensor,

G. Gui, B. Lyu, *Optimization for Wireless Powered Communication Networks*,
SpringerBriefs in Electrical and Computer Engineering,
https://doi.org/10.1007/978-3-030-01021-8_1

which can replenish the energy in time and prolong the lifetime of the sensors. In addition, the solution to the problem of energy shortage of wireless devices can make researchers pay more attention to the optimization of system performance, rather than the goal of minimizing system energy consumption. In addition, EH promotes the recycling of rechargeable batteries or supercapacitors, which is more in line with the current concept of green development. However, there are still some disadvantages in EH based on natural energy sources, among which the stability of energy source is more serious. For example, solar and wind energy, which are affected by natural conditions, cannot provide real-time and stable energy supply to wireless devices.

In recent years, wireless energy transfer (WPT) [1, 8] has attracted extensive attention in both academia and industry, for which radio frequency (RF)-WPT is becoming a hot research topic. Compared with natural energy sources, RF signal is less affected by natural factors such as weather. Hence, it is more stable and is convenient for practical deployment. However, due to the path-loss, the signal energy attenuates greatly with the increase of transmission distance, and the energy harvested by devices farther away from the RF source is usually limited. Therefore, RF-WPT is more suitable for low-power devices, such as RFID and the sensors deployed for IoT [9]. With the development of antenna technology and the improvement of energy efficiency of circuit conversion, the efficiency of WPT has been improved significantly [10]. Therefore, it can be predicted that RF-WPT will take a place in future IoT applications.

In fact, the main purpose of RF signal transmission in wireless communication is the transmission of information, so how to combine wireless information transmission with energy transfer has become a new research topic. Hence, the idea of wireless powered communication (WPC) has been proposed [1, 8, 11]. The recent study about WPC has two directions, i.e., simultaneous wireless information and power transfer (SWIPT) [12–14] and wireless powered communication network (WPCN) [15–18]. Since modulated RF signals also carrying energy, wireless devices can be powered by the energy contained in the received RF signals, i.e., wireless devices can harvest energy and decode information from the RF signals, which is termed SWIPT. Generally, any network consisting of devices wirelessly powered by RF-WPT can be a WPCN [20]. It is known that information transmission of wireless devices includes active communication and passive communication. Recently, the researches about WPCN focus on active communication, where the famous harvest-then-transmit (HTT) mode is adopted [15]. Following the HTT mode, wireless devices first harvest energy from the signals radiated by the RF source and then use the harvested energy for information transmission. While, the passive communication for WPCN typically follows the backscatter communication (BackCom) mode [21–24]. The devices following the BackCom mode backscatter the incident signals for information transmission such that any active components are not required.

A classical WPCN usually includes a hybrid access point (HAP) with a stable energy source and multiple energy-constrained users (wireless devices), which is operated following the HTT mode. The HTT mode consists of wireless power transfer (WPT) phase and wireless information transmission (WIT) phase [15]. The users harvest energy from the RF signals radiated by the HAP during the WPT phase, and the users transmit information using the harvested energy to the HAP during the WIT phase. In [15], a WPCN with a half-duplex (HD) HAP was studied, where the HAP cannot broadcast energy and receive information simultaneously. The authors investigated the tradeoff between the WPT phase and the WIT phase to maximize the sum-throughput and derived the optimal time allocation scheme, from which a doubly near-far problem was revealed. This problem is due to both the downlink (DL) and uplink (UL) distance-dependent signal attenuation, where a far user from the HAP receives less energy than a nearer user in the DL and has to transmit with more power in the UL for reliable information transmission. To overcome this problem, the authors further studied the common-throughput maximization to improve the user fairness. In [16], the authors extended the HD-HAP to a full-duplex (FD)-HAP, which has two antennas, where one antenna is used for broadcasting energy to all users and the other antenna is used for receiving the information from all users via time division multiple access (TDMA). It is obvious that the users' harvesting time is improved significantly since one users can always harvest energy other than the time slot in which the user transmit information. However, the HAP may suffer from the self-interference problem since the HAP broadcasts energy and receive information simultaneously. Fortunately, the self-interference can be cancelled by some successive interference cancellation (SIC)techniques [25–28]. One typical application of WPCN is the wireless sensor network, where the users are usually low-power consumption sensors. These users typically store the harvested energy in supercapacitors, which suffer from the high self-discharge. Hence, the stored energy may be not available for the next communication cycle. To deal with this practical limitation, the authors in [17] considered the energy causality constraint in a WPCN with FD-HAP. In this system model, it was assumed that the users harvest energy before its information transmission but not after. In [18], the authors investigated the joint time and power allocation under energy causality constraints, where both the infinite and finite battery capacity cases are considered. For the infinite battery capacity case, the HAP consumed all energy for WPT during the first several time slots. While for the finite battery capacity case, the authors designed an optimal algorithm to maximize the system throughput.

Non-orthogonal multiple access (NOMA) has been considered as a promising multiple access scheme for future radio access, which can improve system efficacy and meet 5G requirements [29]. The application of NOMA in WPCN has recently attracted a great deal of attention [30–33]. In [30], NOMA was first applied in WPCN. The authors investigated the system throughput maximization problem and analyzed the effect of two decoding order schemes on system performance. The simulation results indicated that NOMA can significantly improve the user fairness. In [31], the authors studied WPCN with NOMA based on time-sharing and derived the optimal time allocation in closed-form to maximize the proportional fairness. In [30] and [31], only the terminals are only with one single antenna. To further improve the system performance, the authors in [32] applied multiple-input multiple output (MIMO) in WPCN with NOMA, where the base station has multiple antennas and the users are all with one single antenna. A sum-throughput maximization problem was studied by joint designing the energy beamforming (EB), receiver beamforming (RB) and time allocation. However, this problem is a NP-hard problem and a two-step method was thus proposed to solve it. In the first step, the successive convex approximation was used to derive the optimal EB and RB with the fixed time allocation. In the second step, based the obtained result derived in the first step, one-dimensional search was used to derive the optimal time allocation. In [33], the authors studied the performance of two decoding schemes in NOMA-WPCN, i.e., the low-complexity decoding scheme and the SIC decoding scheme. Different from [30–32] where the system performance was only studied during one single time slot. The authors in [33] investigated the sum-throughput maximization problem over a finite horizon.

The WPCNs mentioned above are all assumed that each WPCN is with a dedicated spectrum band, so there is no need to consider the interference caused by other communication systems. However, spectrum is a scarce resource and it is impractical to allocate a dedicated spectrum for each WPCN. To tackle this limitation, the authors first integrated cognitive radio (CR) with WPCN, termed cognitive WPCN (CWPCN), where the secondary WPCN share spectrum with the primary communication system for energy and information transmission [34]. The authors considered two systems, i.e., underlay based CWPCN and overlay based CWPCN, and investigated the secondary system throughput maximization problem, the optimal solution may bring user unfairness problem. To improve user fairness, the authors in [35] studied the proportional fair scheduler problem in CWPCN and designed an algorithm to derive the optimal tradeoff between the user fairness and system throughput. The authors applied MIMO in CWPCN, where the HAP and users are all with multiple antennas [36]. To maximize the system throughput, the authors investigated the joint optimization of transmit energy and information covariance matrices and time allocation.

According to the generalized definition of WPCN, backscatter communication system (BCS) can be regarded as a special WPCN. Different from the classical WPCN, the users in a BCS transmit information in the backscatter communication (BackCom) mode. In [24], the authors designed a novel bistatic BCS, where the carrier emitter (CE) is detached from the reader. Under this setup, the communication range can be significantly extended. Ambient backscatter was first proposed in 2013 [21]. Different from the bistatic BCS, in the ambient BCS, the incident signals are from the ambient RF sources such that the dedicated CEs are required. To improve the transmission rate and range, multiple antennas were adopted in BCS. The authors in [37] investigated the signal detection problem in ambient BCS, for which the differential encoding was adopted to eliminate the necessity of channel estimation.

Inspired by the advantages of the HTT and BackCom modes, integrating the HTT and BackCom modes in WPCN has recently a lot of attention. Ambient backscatter (AB) was first applied in overlay-CWPCN in [38], which includes a primary transmitter (PT), a secondary user (CU) and a secondary receiver (SR), where the CU can work in the AB mode and the HTT mode. When the PT is activated, the CU backscatters information to the SR or harvests energy for further use. When the PT is idle, the CU adopt the HTT mode for information transmission. Under this setup, the authors designed the optimal time allocation scheme to maximize the system throughput. The authors in [39] extended the work in [38] to the underlay-CWPCN, where the PT is always activated during a given block. From the optimal time scheme for this scenario, the authors found that the CU works in either the AB mode or the HTT mode but not both. The authors in [40] further considered a multi-CUs scenario and showed that the increase of the number of the CUs can improve the system performance. The authors in [41] considered a hybrid network, where the PT and the power beacon (PB) both serve as the energy source. If CU is in the PB coverage, the CU can work in both the AB mode and the bistatic backscatter (BB) mode; otherwise, the CU can only adopt the AB mode. It is shown that the hybrid BackCom mode can improve the system throughput.

1.2 Basic Modes of Wireless Powered Communication Networks

In this section, we introduce the basic principals of the HTT and BackCom modes in detail.

1.2.1 HTT Mode

As stated in [15], the HTT mode consist of two phase, i.e., the WPT phase and WIT phase. During the WPT phase, the energy source broadcasts RF signals to users and the users convert the received signals into direct current using a rectifier and store it in their energy storages. During the WIT phase, the users use the harvested energy to transmit information. In a WPCN, there are two scenarios in which the energy transmitter (ET) and the information receiver (IR) are co-located and separately-located, as shown in Figs. 1.1 and 1.2, respectively. For the co-located scenario, the ET and IR constitute an HAP, which works in the HD mode or the FD mode depending the number of its antenna(s). Generally, if the HAP only has one single antenna, it adopts the HD mode; while if the the HAP has two antennas, it can work in the FD mode. For the separately-located scenario, the HAP can work in the HD mode or the FD mode.

Fig. 1.1 ET and IR are co-located [15–17]. (**a**) HD-HAP. (**b**) FD-HAP

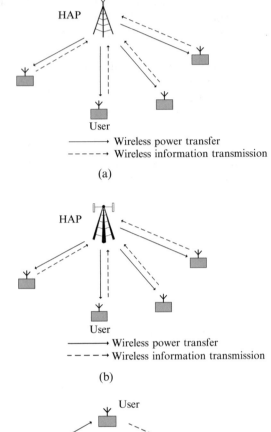

(a)

(b)

Fig. 1.2 ET and IR are co-located [19]

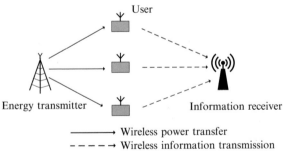

1.2.1.1 HTT Mode Based on HD-HAP

We introduce the HTT mode based on an HD-HAP following the system model shown in Fig. 1.1a. At present, multiple users in WPCN usually transmit information in multiple access modes such as TDMA or NONA. The time allocation diagram of the HTT mode based on an HD-HAP under TDMA and NOMA is shown in Fig. 1.3. In Fig. 1.3a, if K users transmit information via TDMA, the duration of the WIT phase is divided into K time slots. Each user transmit information during its

Fig. 1.3 The time allocation diagram of the HTT mode based on an HD-HAP [15, 30]. (**a**) TDMA scheme. (**b**) NOMA scheme

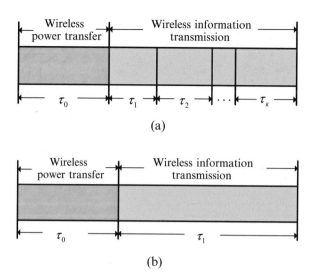

(a)

(b)

allocated time slot to avoid inferences, which also causes user unfairness problem. If NOMA is employed, all users transmit information to the HAP simultaneously and the HAP decodes each user's information by SIC, which can improve user fairness but also make the demodulation system at the HAP more complex.

1.2.1.2 HTT Mode Based on FD-HAP

If the HAP has two antennas, it can adopt the FD mode for transmitting energy and receiving information simultaneously. We introduce the HTT mode based on an FD-HAP following the system model shown in Fig. 1.1b. Depending on whether or not energy causality constraints are taken into account, the time allocation diagrams corresponding to the HTT mode based on an FD-HAP are shown in Fig. 1.4a, b, respectively. In Fig. 1.4a, the transmission block is divided into $K + 1$ time slots. Since the discharge characteristic is not considered, each user can harvest energy in K time slots other than the time slot in which the user transmits information. For example, the energy harvesting time of user-i is $\sum_{j=\neq i}^{K} \tau_j$. However, the energy harvesting time of user-i following the HTT mode based on an HD-HAP is τ_i. It is obvious that the HTT mode based on an FD-HAP can significantly the energy harvesting time. The users in WPCN usually use the supercapacitors to store energy, which are with the high discharge characteristic. Hence, the stored energy time of these users are typically limited and the energy causality constraints need to be considered. Following the energy causality constraints, it is assumed that the users can only used the energy harvested before its information transmission but not after. The time allocation diagram is shown in Fig. 1.4b. Under the energy causality constraints, the users' harvesting energy time will reduce, especially for those who are scheduled first.

Fig. 1.4 The time allocation diagram of the HTT mode based on an FD-HAP [16, 17]. (**a**) Without energy causality constraints. (**b**) With energy causality constraints

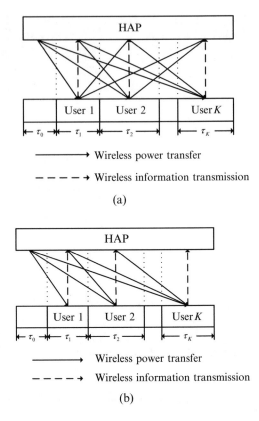

Since the HAP simultaneously transmit energy and receive information through the same band, it also suffers from the self-interference. Fortunately, the energy signals as the interference is known to the HAP, so the interference can be eliminated. Recently, digital cancellation and analog cancellation are two main interference cancellation techniques, which can cancel the interference to the noise level. Hence, the HTT mode based on an FD-HAP is also practical.

1.2.2 BackCom Mode

Backscatter communication has been extensively used in low-power sensors network. Different from the conventional communication systems, backscatter communication does not need any active component, and the user (tag) reflects the incident signals for information transmission by adjusting the mismatch of it antenna impedance. According to [42], the user structure is illustrated in Fig. 1.5. From Fig. 1.5 the micro-controller achieves load modulation via changing the load

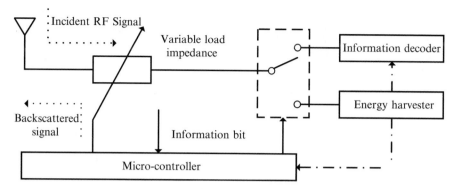

Fig. 1.5 The structure of a BCS User [42]

impedance based on the pre-determined signal bits. Denote the load impedance and the antenna impedance as Z_i and Z_a, respectively. We use on-off keying to illustrate the user's information modulation in BCS. The reflection coefficient is defined as

$$\Gamma_i = \frac{Z_i - Z_a^*}{Z_i + Z_a},$$ (1.1)

where $*$ denotes the complex conjugate operation. The reflection coefficient is changed by switching the load impedance between Z_0 and Z_1. If $Z_i = Z_a^*$, the antenna impedance is matched and all incident signals will be absorbed, which corresponds to bit "0". If $Z_i \neq Z_a^*$, the antenna impedance is dismatched and part of incident signals will be reflected, which corresponds to bit "1". The discussions about higher-order modulation for BCS can refer to [23].

Recently, the BackCom mode can be classified into three types, i.e., monostatic backscatter (MB), BB and AB. We briefly introduce the three types as follows according to [43].

1.2.2.1   Monostatic BCS

As shown in Fig. 1.6a, the system consists of a reader (Interrogator) and a user (Tag). In this system, the reader first serves as the CE to transmit carrier signals to active the user. Once the user activated, it modulates and reflects signals via adjust the mismatch of the antenna impedance, while the reader serves as the IR to receive the backscattered signals. Since the CE and IR are integrated in the reader, the carries signals suffer from the double path-loss attenuation. Generally, the monostatic BCS is suitable for short-range RFID applications.

Fig. 1.6 The diagram of
backscatter communication
systems. (**a**) Monostatic BCS.
(**b**) Bistatic BCS. (**c**) Ambient
BCS[43]

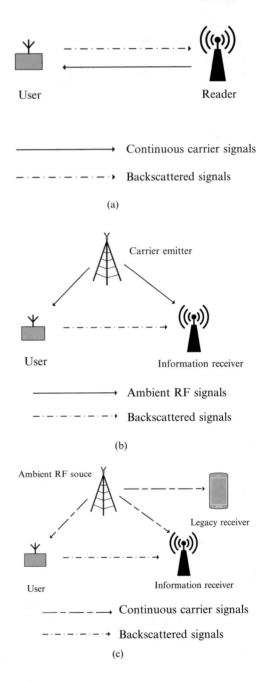

1.2.2.2 Bistatic BCS

Different from the monostatic BCS, the CE and IR are separately located, which is shown in Fig. 1.6b. It is obvious that the double path-loss attenuation problem can be avoided. Compared with the monostatic BCS, this system can build more flexible network topologies. For example, if the locations of the user and the IR are fixed, the optimal location of the CE can be design to improve the system performance. Moreover, multiple CEs can be deployed around the user such that the communication range can be extended.

1.2.2.3 Ambient BCS

Similar as the bistatic BCS, the CE and IR are also separately located. However, in ambient BCS, the CE is not dedicated. Instead, the CE can be ambient RF sources such as WiFi access points and TV towers. The user can reflect the received RF signals from ambient environment for information backscattering, which is shown in Fig. 1.6c. However, since the ambient signals cannot be controlled, the system performance may be not stable.

References

1. S. Bi, C. K. Ho, and R. Zhang, "Wireless powered communication: opportunities and challenges," *IEEE Communications Magazine*, vol. 53, no. 4, pp. 117–125, Apr. 2015.
2. S. Bi, Y. Zeng, and R. Zhang, "Wireless powered communication networks: an overview," *IEEE Wireless Communications*, vol. 23 no. 2, pp. 10–18, Apr. 2016.
3. C. K. Ho and R. Zhang, "Optimal energy allocation for wireless communications with energy harvesting constraints," *IEEE Transactions on Signal Processing*, vol. 60, no. 9, pp. 4808–4818, Sept. 2012.
4. C. Huang, R. Zhang, and S. Cui, "Optimal power allocation for outage probability minimization in fading channels with energy harvesting constraints," *IEEE Transactions on Wireless Communications*, vol. 13, no. 2, pp. 1074–1087, Feb. 2014.
5. V. Raghunathan, A. Kansal, J. Hsu, J. Friedman, and M. Srivastava, "Design considerations for solar energy harvesting wireless embedded systems," in *Proc. IPSN*, pp. 457–462, LA, California, USA, 2005.
6. Y. K. Tan and S. K. Panda, "Optimized wind energy harvesting system using resistance emulator and active rectifier for wireless sensor nodes," *IEEE Transactions on Power Electronics*, vol. 26, no. 1, pp. 38–50, Jan. 2011.
7. X. Lu and S-H. Yan, "Thermal Energy Harvesting for WSNs," in *Proc. IEEE SMC*, pp. 3045–3052, Istanbul, Turkey, Oct. 2010.
8. K. Huang, C. Zhong, and G. Zhu, "Some new research trends in wirelessly powered communications," *IEEE Wireless Communications*, vol. 23, no. 2, pp. 19–27, Apr. 2016.
9. P. Ramezani and A. Jamalipour, "Toward the evolution of wireless powered communication networks for the future Internet of Things," IEEE Network, vol. 31, no. 6, pp. 62–69, Nov. /Dec. 2017.
10. Y. Suh and K. Chang, "A high-efficiency dual-frequency rectenna for 2.45- and 5.8-GHz wireless power transmission," *IEEE Transactions on Microwave Theory and Techniques*, vol. 50, no. 7, pp. 1784–1789, Jul. 2002.

11. K. Huang and X. Zhou, "Cutting the last wires for mobile communications by microwave power transfer," *IEEE Communications Magazine*, vol. 53, no. 6, pp. 86–93, June 2015.
12. R. Zhang and C. K. Ho, " MIMO broadcasting for simultaneous wireless information and power transfer.," *IEEE Transactions on Wireless Communications*, vol. 12, no. 5, pp. 1989–2001, May 2013.
13. L. Liu, R. Zhang, and K. C. Chua, "Wireless information and power transfer: A dynamic power splitting approach," *IEEE Transactions on Communications*, vol. 61, no. 9, pp. 3990–4001, Sept. 2013.
14. H. Xing, L. Liu, and R. Zhang, "Secrecy wireless information and power transfer in fading wiretap channel," *IEEE Transactions on Vehicular Technology*, vol. 65, no. 1, pp. 180–190, Jan. 2016.
15. H. Ju and R. Zhang, "Throughput maximization in wireless powered communication networks," *IEEE Transactions on Wireless Communications*, vol. 13, no. 1, pp. 418–428, Jan. 2014.
16. H. Ju and R. Zhang, "Optimal resource allocation in full-duplex wireless-powered communication network," *IEEE Transactions on Communications*, vol. 62, no. 10, pp. 3528–3540, Oct. 2014.
17. X. Kang, C. K. Ho, and S. Sun, "Full-duplex wireless-powered communication network with energy causality," *IEEE Transactions on Wireless Communications*, vol. 14, no. 10, pp. 5539–5551, Oct. 2015.
18. H. Lee, K. J. Lee, H. Kim, B. Clerckx, and I. Lee, "Resource allocation techniques for wireless powered communication networks with energy storage constraint," *IEEE Transactions on Wireless Communications*, vol. 15, no. 4, pp. 2619–2628, Apr. 2016.
19. Q. Wu, W. Chen, and J. Li, "Wireless powered communications with initial energy: Qos guaranteed energy-efficient resource allocation," *IEEE Communications Letters*, vol. 19, no. 12, pp. 2278–2281, Dec. 2015.
20. P. Ramezani, "Extending Wireless Powered Communication Networks for Future Internet of Things," Australia: Faculty of Engineering and Information Technologies, University of Sydney.
21. V. Liu, A. Parks, V. Talla, S. Gollakota, D. Wetherall, and J. R. Smith, "Ambient backscatter: Wireless communication out of thin air," in *Proc. ACM SIGCOMM*, pp. 39–50, Hong Kong, Aug. 2013.
22. B. Kellogg, A. Parks, S. Gollakota, J. R. Smith, and D. Wetherall, "WiFi backscatter: Internet connectivity for RF-powered devices," in *Proc. ACM SIGCOMM*, pp. 607–608, New York, NY, USA, 2014.
23. C. Boyer and S. Roy, "Backscatter communication and RFID: Coding, energy, and MIMO analysis," *IEEE Transactions on Communications*, vol. 62, no. 3, pp. 770–785, Mar. 2014.
24. J. Kimionis, A. Bletsas, and J. N. Sahalos, "Increased range bistatic scatter radio," *IEEE Transactions on Communications*, vol. 62, no. 3, pp. 1091–1104, Mar. 2014.
25. J. I. Choi, M. Jain, K. Srivivasan, P. Levis, and S. Katti, "Achieving single channel, full duplex wireless communication," in *Proc. ACM MobiCom*, pp. 1–12, Illinois, USA, Sept. 2010.
26. D. Bharadia, E. McMilin, and S. Katti. Full duplex radios, in *ACM SIGCOMM*, pp. 375–386, Hong Kong, China, Aug. 2013.
27. B. P. Day, A. R. Margetts, D. W. Bliss, and P. Schniter, "Full-duplex bidirectional MIMO: Achievable rates under limited dynamic range," *IEEE Transactions on Signal Processing*, vol. 60, no. 7, pp. 3702–3713, Jul. 2012.
28. T. M. Kim, H. J. Yang, and A. J. Paulraj, "Distributed sum-rate optimization for full-duplex MIMO system under limited dynamic range," *IEEE Signal Processing Letter*, vol. 20, no. 6, pp. 555–558, Jun. 2013.
29. L. Dai, B. Wang, Y. Yuan, S. Han, C. l. I, and Z. Wang, "Non-orthogonal multiple access for 5G: Solutions, challenges, opportunities, and future research trends," *IEEE Communications Magazine*, vol. 53, no. 9, pp. 74–81, Sept. 2015.
30. P. D. Diamantoulakis, K. N. Pappi, Z. Ding and G. K. Karagiannidis, "Wireless-powered communications with non-orthogonal multiple access," *IEEE Transactions on Wireless Communications*, vol. 15, no. 12, pp. 8422–8436, Dec. 2016.

31. P. D. and G. K. Karagiannidis, "Maximizing proportional fairness in wireless powered communications," *IEEE Wireless Communications Letters*, vol. 6, no. 2, pp. 202–205, Apr. 2017.

32. Y. Yuan and Z. Ding, "The application of non-orthogonal multiple access in wireless powered communication networks," in *Proc. IEEE SPAWC*, pp. 1–5, Edinburgh, UK, 2016.

33. M. A. Abd-Elmagid, A. Biason, T. ElBatt, K. G. Seddik, and M. Zorzi, "Non-orthogonal multiple access schemes in wireless powered communication networks," in *Proc. ICC*, pp. 1–6, Paris, France, 2017.

34. S. Lee and R. Zhang, "Cognitive wireless powered network: Spectrum sharing models and throughput maximization," *IEEE Transactions on Cognitive Communications and Networking*, vol. 1, no. 3, pp. 335–346, Sept. 2015.

35. Y. Cheng, P. Fu, Y. Ding, B. Li, and X. Yuan, "Proportional fairness in cognitive wireless powered communication networks," *IEEE Communications Letters*, vol. 21, no. 6, pp. 1397–1400, June 2017.

36. J. Kim, H. Lee, C. Song, T. Oh, and I. Lee, "Sum throughput maximization for multi-user MIMO cognitive wireless powered communication networks," *IEEE Transactions on Wireless Communications*, vol. 16, no. 2, pp. 913–923, Feb. 2017.

37. G. Wang, F. Gao, R. Fan, and C. Tellambura, " Ambient backscatter communication systems: Detection and performance analysis," *IEEE Transactions on Communications*, vol. 64, no. 11, pp. 4836–4846, Nov. 2016.

38. D. T. Hoang, D. Niyato, P. Wang, D. I. Kim, and Z. Han, " The tradeoff analysis in RF-powered backscatter cognitive radio networks," in *Proc. IEEE GLOBECOM*, pp. 1–6, Washington, DC, USA, Dec. 2016.

39. D. T. Hoang, D. Niyato, P. Wang, D. I. Kim, and Z. Han, "Ambient backscatter: A new approach toimprove network performance for RF-powered cognitive radio networks," *IEEE Transactions on Communications*, vol. 65, no. 9, pp. 3659–3674, Sept. 2017.

40. D. T. Hoang, D. Niyato, P. Wang, and D. I. Kim, "Optimal time sharing in RF-powered backscatter cognitive radio networks," in *Proc. IEEE ICC*, pp. 1–6, Paris, France, May 2017.

41. S. H. Kim and D. I. Kim, "Hybrid backscatter communication for wireless-powered heterogeneous networks," *IEEE Transactions on Wireless Communications*, vol. 16, no. 10, pp. 6557–6570, Oct. 2017.

42. S. Gong, X. Huang, J. Xu, W. Liu, P. Wang, and D. Niyato, "Backscatter relay communications powered by wireless energy beamforming," *IEEE Transactions on Communications*, https://doi.org/10.1109/TCOMM.2018.2809613.

43. N. V. Huynh, D. T. Hoang, X. Lu, D. Niyato, P. Wang, and D. I. Kim, "Ambient backscatter communication: A contemporary survey," arXiv preprint: 1712.04804, 2017.

Chapter 2
Non-orthogonal Multiple Access in Wireless Powered Communication Networks

Abstract In a WPCN, the distances between the users and the power beacon (PB) or the information receiver (IR) are different, which results in unfair transmission rates among different users. NOMA, the basic principle of which is that the users can achieve multiple access by exploiting the power domain multiplexing, is applied in the WPCN to improve user fairness. However, the successive interference cancellation (SIC) constraints, the prerequisite of applying NOMA successfully, are not considered in most existing literatures. In this chapter, the effect of SIC constraints on the throughput of the WPCN with NOMA is investigated, where the users harvest energy from RF signals radiated by the PB, and then use the harvested energy to simultaneously transmit information to the IR. First, the throughput maximization problem is formulated to find the optimal time and energy allocation scheme. Then, to derive the closed-form solution, the optimization problem is further divided into two sub-problems by exploiting the optimal structure of constraints. Finally, simulations on the effect of SIC constraints show the importance of the distinctness among users' channel power gains for the WPCN with NOMA.

2.1 Introduction

As a key technology in the future wireless communication, non-orthogonal multiple access (NOMA) has attracted more and more attention of researchers [2, 3]. Different from the orthogonal multiple access methods such as TDMA and FDMA, employing NOMA, users can simultaneously transmit information in the same frequency and realize multiple access in power domain, which can effectively improve the spectrum efficiency and improve the communication performance of users with poor channel conditions. In WPCNs, the transmission efficiencies of users are different due to the distances between users and the power beacon (PB) are different, which indicates the users with better channel conditions can harvest more energy and transmit more information during a given time slot, resulting in unfairness between users. In other words, the orthogonal multiple access methods may be not the best choice for WPCN from the angle of avoiding unfairness between users [4]. If users can transmit information simultaneously via NOMA [4–7], the throughput of users with poor channel conditions will be improved effectively and the fairness between users will be improved. In NOMA-WPCNs, successive interference cancellation (SIC) is widely used to decode information from multiple users. For efficient SIC at the information receiver (IR), the SIC constraint at the IR should be satisfied [8]. The SIC constraint defined as that the power difference between the signal to be decoded and the remaining signals should be greater than a given threshold.

However, the SIC constraint is ignored in most of the literature on NOMA-WPCN [4–7]. Therefore, it is of great significance to study the effect of SIC constraints in NOMA-WPCN on system performance for the practical application of NOMA in WPCN. In this chapter, we consider a practical WPCN with NOMA by considering the SIC constraints, where the power station and information receiver are separately located. The main contributions of this chapter can be summarized as follows. First, we consider a practical WPCN with NOMA under SIC constraints, where the users first harvest energy from the PB and then transmit information to the IR via NOMA. Second, we formulate an optimization problem to maximize the system throughput, which is non-convex. It is then shown that the non-convex optimization problem can be transformed into a convex optimization problem by jointly optimizing the time allocation of the HTT protocol and the consumed energy at the users, which can be further divided into two sub-problems by exploiting the optimal structure of the constraints. Then, we derive a closed-form solution by solving the two sub-problems respectively. Third, the simulation results are conducted to show the effect of SIC constraints on the system performance and the importance of the distinctness among users' channels in the proposed model. Note that the majority of the contents of this chapter are based on our previous work [1].

2.2 System Model

We consider a WPCN, which consists of a PB, an IR, and two users with batteries. We assume that the PB has a stable energy supply, and the users denoted as $U_i, i = 1, 2$ do not have a energy source and need to harvest energy from the PB. All terminals are equipped with one single antenna. Denote the channel power gains from the PB to U_i and from U_i to the IR as h_i and g_i. We assume that h_i and g_i are are quasi-static flat fading and remain constant during one block but may change from one block to another. It is assumed that the network has the prefect information about the channel power gains. The system model is shown in Fig. 2.1. We study the system performance during a transmission block, of which the duration is T seconds.

In this network, the HTT mode is adopted. Hence, the transmission block has two phase, i.e., wireless power transfer (WPT) phase and wireless information transmission (WIT) phase. During the WPT phase, the users harvest energy radiated by the PB and store the harvested energy in their batteries. During the WIT phase,

Fig. 2.1 Model of a WPCN with NOMA

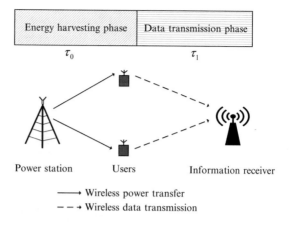

all users transmit independent information using the harvested energy to the IR simultaneously via NOMA. The SIC technique is adopted at the IR such that the IR can decode the users' information respectively. Denote the duration of each phase as τ_0 and τ_1, respectively, and we have the *time constraint* $\tau_0 + \tau_1 \leq T$. Denote the transmit power of the PB as P. During the WPT phase, the harvested energy of U_i is given by

$$\hat{E}_i = \eta_i P h_i \tau_0, \quad i = 1, 2$$

where η_i is the energy harvesting efficiency of U_i. Without loss of generality, we assume $\eta = \eta_1 = \eta_2$. Following the assumption in [9], we know that the main power consumption during the WIT phase is for information transmission. Hence, we do not consider the circuit energy consumption for simplicity. Denote the transmit power of U_i as P_i, which satisfies the *energy constraints*

$$P_i \tau_1 \leq \hat{E}_i, \quad i = 1, 2.$$

Since the users transmit information via NOMA, the IR simultaneously receives the information from all users. According to [4], the indices of the two users are assigned in a way that $h_i g_i$ are sorted in descending order, i.e., $h_1 g_1 \geq h_2 g_2$ [4]. To decode each user's message, the following *SIC constraints* should be satisfied by controlling each user's transmit power, which are given by

$$P_1 \gamma_1 - P_2 \gamma_2 \geq P_{\text{th}}, \tag{2.1}$$

where $\gamma_i = \frac{g_i}{\sigma^2}$, σ^2 is the noise power at the information receiver, P_{th} is the minimum power difference required to distinguish between the signals to be decoded and the remaining non-decoded signals (named threshold power for simplicity) [8]. The illustration of uplink NOMA with SIC at the IR is given in Fig. 2.2. From Fig. 2.2, we know that if $P_1 \gamma_1 - P_2 \gamma_2 \geq P_{\text{th}}$, the information receiver first decodes the

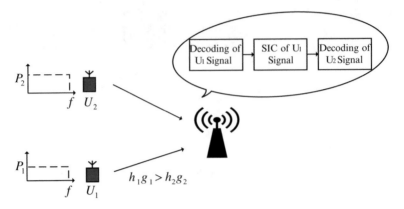

Fig. 2.2 Illustration of a two-user uplink NOMA with SIC at the information receiver

signal of U_1 by treating the signal of U_2 as noise. Then, the information receiver subtracts the decoded signal of U_1 from the received signal and decodes the signal of U_2 [3]. Note that if the SIC constraints are not satisfied, the SIC technique cannot be efficiently applied to decode signals at the information receiver [8]. With the satisfaction of the SIC constraints, the signal to interference plus noise ratio (SINR) of U_i at the information receiver is expressed as

$$\Gamma_1 = \frac{P_1 g_1}{P_2 g_2 + \sigma^2} = \frac{P_1 \gamma_1}{P_2 \gamma_2 + 1},$$

$$\Gamma_2 = P_2 \gamma_2.$$

Hence, the achievable throughput of U_i can be formulated as

$$R_1 = \tau_1 \log_2(1 + \Gamma_1) = \tau_1 \log_2(1 + \frac{P_1 \gamma_1}{P_2 \gamma_2 + 1}), \tag{2.2}$$

$$R_2 = \tau_1 \log_2(1 + P_2 \gamma_2). \tag{2.3}$$

Hence, the total sum-throughput of the system is given as

$$R_{\text{sum}} = \sum_{i=1}^{2} R_i = \tau_1 \log_2(1 + \sum_{i=1}^{2} P_i \gamma_i) - \tau_1 \log_2(1 + P_2 \gamma_2)$$

$$+ \tau_1 \log_2(1 + P_2 \gamma_2) = \tau_1 \log_2(1 + \sum_{i=1}^{2} P_i \gamma_i). \tag{2.4}$$

2.3   Sum-Throughput Maximization

In this section, we aim to maximize the sum-throughput with the joint optimization of both time and energy allocation. First, the sum-throughput maximization problem is formulated as follows

$$\max_{\tau, P} \quad R_{\text{sum}}$$

$$\text{s.t.} \quad \text{C1: } P_i \tau_1 \leq \eta P h_i \tau_0, \quad i = 1, \cdots, 2,$$

$$\text{C2: } P_1 \gamma_1 - P_2 \gamma_2 \geq P_{\text{th}},$$

$$\text{C3: } P_i \geq 0, \quad i = 1, \cdots, 2, \quad (2.5)$$

$$\text{C4: } \tau_0, \tau_1 \geq 0,$$

$$\text{C5: } \tau_0 + \tau_1 \leq T.$$

where $\tau = [\tau_0, \tau_1]$, $P = [P_1, P_2]$. Assume that (2.5) is feasible. Equation (2.5) is a non-convex optimization problem due to the coupling of P_i and τ_1 in the objective function and energy constraints. We introduce auxiliary variables $E_i = P_i \tau_1$ for $i = 1, 2$ in (2.5). With the new variables, R_{sum} is reformulated as

$$R_{\text{sum}}(\tau, E) = \tau_1 \log_2 \left(1 + \frac{\sum_{i=1}^{2} E_i \gamma_i}{\tau_1}\right), \quad (2.6)$$

and the constraints C1–C3 are rewritten as

$$\text{C6:} \quad E_i \leq \eta P h_i \tau_0, \quad i = 1, 2,$$

$$\text{C7:} \quad E_1 \gamma_1 - E_2 \gamma_2 \geq P_{\text{th}} \tau_1,$$

$$\text{C8:} \quad E_i \geq 0, \quad i = 1, 2.$$

Denote $E = [E_1, \cdots, E_2]$. Equation (2.5) is reformulated as (2.7), which is given as follows.

$$\max_{\tau, E} \quad \tau_1 \log_2 \left(1 + \frac{\sum_{i=1}^{2} E_i \gamma_i}{\tau_1}\right)$$

$$\text{s.t.} \quad \text{C4, C5, C6, C7, C8.} \quad (2.7)$$

Lemma 2.1 *Equation (2.7) is a convex optimization problem.*

Proof Define $f_1(E) = \log_2(1 + \sum_{i=1}^{2} E_i \gamma_i)$. It is easy to prove that $f_1(E)$ is a concave function since $f_2(E_i) = \log_2(1 + E_i \gamma_i)$ is a concave function. Note that $R_{\text{sum}}(\tau, E)$ is a perspective function of $f_1(E)$. Due to that the perspective operation preserves convexity, $R_{sum}(\tau, E)$ is thus a concave function [10]. Moreover, all constraints in (2.7) are affine. Hence, (2.7) is a convex optimization problem. This completes Lemma 2.1.

Denote the optimal solution for (2.7) as $E^* = [E_1^*, E_2^*]$ and $\tau = [\tau_0^*, \tau_1^*]$.

Lemma 2.2 *In the optimal condition, $\tau_0^* + \tau_1^* = T$.*

Proof Define the optimal solution for (2.7) as $\{\bar{\tau}, \bar{E}\}$, where $\bar{\tau} = [\bar{\tau}_0, \bar{\tau}_1]$, $\bar{E} = [\bar{E}_1, \bar{E}_2]$, and $\bar{\tau}_0 + \bar{\tau}_1 < T$. We will show that $\bar{\tau}$ is not the optimal solution by contradiction. Assume there are τ^* and E^* satisfying the following conditions:

$$\tau_0^* + \tau_1^* = T, \quad \tau_0^* > \bar{\tau}_0, \quad \tau_1^* = \bar{\tau}_1, \quad \text{and } E_2^* = \bar{E}_2,$$

Let $E_1^* = \eta P h_1 \tau_0^* > \bar{E}_1$. Note that τ^* and E^* satisfy all constraints in (2.7). It can be proved that R_{sum} is an increasing function with E_i such that $R_{sum}(\tau^*, E^*) > R_{sum}(\bar{\tau}, \bar{E})$. This contradicts with the assumption that $\{\bar{\tau}, \bar{E}\}$ is the optimal solution. This thus completes the proof.

To derive a closed-form solution for (2.7), we exploit the optimal structures of C6 and C7., with which the optimization problem is first decomposed into two sub-problems. The solutions for the sub-problems can be derived in closed-form based on KKT optimality conditions. Comparing the results of the two sub-problems, we finally derive the optimal solution. From (2.7) and based on Lemma 2.2, the optimization problem for a two-user case is expressed as

$$\max_{\tau_1, E} \quad \tau_1 \log_2(1 + \frac{E_1\gamma_1 + E_2\gamma_2}{\tau_1})$$

$$\text{s.t.} \quad E_i + \eta P h_i \tau_1 \leq \eta P h_i T, \quad i = 1, 2,$$

$$E_1\gamma_1 - E_2\gamma_2 \geq P_{th}\tau_1, \tag{2.8}$$

$$0 \leq \tau_1 \leq T,$$

$$E_i \geq 0, \quad i = 1, 2.$$

Lemma 2.3 *In the optimal condition,*

$$E_1^* + \eta P h_1 \tau_1^* = \eta P h_1 T, \tag{2.9}$$

$$E_2^* + \eta P h_2 \tau_1^* \leq \eta P h_2 T, \tag{2.10}$$

$$E_1^*\gamma_1 - E_2^*\gamma_2 \geq P_{th}\tau_1^*, \tag{2.11}$$

where at least one of the latter two constraints is an equality.

Proof First, it is obvious that the energy constraint for $i = 1$ is an equality in the optimal condition, which is described in (2.9). Otherwise, we can always increase E_1 to achieve a larger throughput without conflicting other constraints. Then, we prove that at least one of (2.10) and (2.11) is an equality in the optimal condition. Define the optimal solution for (2.8) as $\{\bar{\tau}_1, \bar{E}_1, \bar{E}_2\}$, which satisfies

$$\bar{E}_2 + \eta Ph_2 \bar{\tau}_1 < \eta Ph_2 T, \tag{2.12}$$

$$\bar{E}_1 \gamma_1 - \bar{E}_2 \gamma_2 > P_{th} \bar{\tau}_1. \tag{2.13}$$

We will show that $\{\bar{\tau}_1, \bar{E}_1, \bar{E}_2\}$ is not the optimal solution. Assume there is $\{\tau_1^*, E_1^*, E_2^*\}$ satisfying the following conditions $E_1^* = \bar{E}_1$, $\tau_1^* = \bar{\tau}_1$ and $E_2^* = \min\{\eta Ph_2 T - \eta Ph_2 \tau_1^*, \frac{E_1^* \gamma_1 - P_{th}\tau_1^*}{\gamma_2}\}$. Since the objective function of Problem P3 is an increasing function of E_2 such that the Problem P3 with $\{\tau_1^*, E_1^*, E_2^*\}$ can provide a larger throughput than that with $\{\bar{\tau}_1, \bar{E}_1, \bar{E}_2\}$, which contradicts with the assumption that $\{\bar{\tau}_1, \bar{E}_1, \bar{E}_2\}$ is the optimal solution. This thus completes the proof.

Remark 2.1 From (2.9), we conclude that U_1 corresponding to the largest product of channel power gains can always transmit data with using up its harvested energy. The reason is given as follows. In the uplink NOMA system, U_1 does not make any interference to U_2 such that U_1 exhausts its harvested energy to transmit data can significantly improve the system throughput [8]. The conclusion can be extended to other multi-user cases.

Based on Lemma 2.3, (2.8) can be solved by considering two sub-problems. First, we solve the sub-problem under the constrains that both (2.9) and (2.10) are equalities. If (2.9) and (2.10) are equalities, it is easily derived that $h_1\gamma_1 > h_2\gamma_2$ such that the constraint $E_1^* \gamma_1 - E_2^* \gamma_2 \geq P_{th}\tau_1$ can be satisfied. For this sub-problem, R_{sum} is rewritten as $R_{sum}(\tau_1) = \tau_1 \log_2[1 + \frac{aT}{\tau_1} - a]$, where $a = \eta Ph_1\gamma_1 + \eta Ph_2\gamma_2$. Now R_{sum} is a function only with respect to τ_1. From (2.11) and $0 \leq \tau_1 \leq T$, we derive that $0 \leq \tau_1 \leq \frac{b}{b+P_{th}}T$, where $b = \eta Ph_1\gamma_1 - \eta Ph_2\gamma_2$. Hence, (2.8) is rewritten as

$$\max_{\tau_1, E} \quad \tau_1 \log_2[1 + \frac{aT}{\tau_1} - a]$$

$$\text{s.t.} \quad 0 \leq \tau_1 \leq \frac{b}{b + P_{th}} T. \tag{2.14}$$

It is easy to prove that (2.14) is a convex optimization problem according to the perspective operation [10].

Theorem 2.1 *The optimal time control for* (2.14) *is given as*

$$\tau_1^* = \min\left\{ \frac{aT}{z^* - 1 + a}, \frac{bT}{b + P_{th}} \right\}, \tag{2.15}$$

where $z^* > 1$ is the unique solution of $f(z) = a$, and $f(z) = z \ln z - z + 1$.

Proof Since Problem (2.14) is a convex optimization problem, the partial Lagrange function is defined as $\mathscr{L}(\tau_1) = R_{\text{sum}}(\tau_1)$. By applying KKT conditions, we derive that

$$\frac{\partial L}{\partial \tau_1} = \log_2(1 + \frac{aT}{\tau_1^*} - a) - \frac{\frac{aT}{\tau_1^*}}{\ln 2(1 + \frac{aT}{\tau_1^*} - a)}$$

$$\begin{cases} = 0, & 0 < \tau_1^* \le \frac{bT}{b+P_{\text{th}}}, \\ > 0, & \tau_1^* = \frac{bT}{b+P_{\text{th}}}. \end{cases} \tag{2.16}$$

If $0 < \tau_1^* \le \frac{bT}{b+P_{\text{th}}}$, we can derive that

$$z\ln z - z + 1 = a, \tag{2.17}$$

where $z = 1 + \frac{aT}{\tau_1^*} - a$. It can be found that $z > 1$ since $a > 0$ and $0 < \tau_1^* \le \frac{bT}{b+P_{\text{th}}}$. From Lemma 3.2 in [9], we can find a unique $z^* > 1$ satisfying 2.17. From $z^* = 1 + \frac{aT}{\tau_1^*} - a$, we derive that $\tau_1^* = \frac{aT}{z^*-1+a}$. Combing $\tau_1^* \le \frac{bT}{b+P_{\text{th}}}$, we derive the result as shown in Theorem 2.1.

Based on the conditions that (2.9) and (2.10) are equalities, the optimal energy allocation for (2.8) is given by

$$E_1^* = \eta P h_1(T - \tau_1^*), \tag{2.18}$$

$$E_2^* = \eta P h_2(T - \tau_1^*). \tag{2.19}$$

The other sub-problem is under the constraints that both (2.9) and (2.11) are equalities. For this sub-problem, $R_{\text{sum}}(\tau_1) = \tau_1 \log_2(1 + \frac{2cT}{\tau_1} - d)$, where $c = \eta P h_1 \gamma_1$ and $d = 2\eta P h_1 \gamma_1 + P_{\text{th}}$. Since $E_2 \ge 0$, we introduce (2.9) into (2.11) and derive that $\tau_1 \le \frac{cT}{c+P_{\text{th}}}$. In addition, we introduce (2.9) and (2.11) into (2.10) and derive $\frac{bT}{b+P_{\text{th}}} \le \tau_1$. Since $b < c$, we derive that $\frac{bT}{b+P_{\text{th}}} < \frac{cT}{c+P_{\text{th}}}$. Finally, the new constraint of this sub-problem is given as $\frac{bT}{b+P_{\text{th}}} \le \tau_1 \le \frac{cT}{c+P_{\text{th}}}$. Based on the above analysis, (2.8) is rewritten as

$$\max_{\tau_1, E} \quad \tau_1 \log_2(1 + \frac{2cT}{\tau_1} - d)$$

$$\text{s.t.} \quad \frac{bT}{b + P_{\text{th}}} \le \tau_1 \le \frac{cT}{c + P_{\text{th}}}. \tag{2.20}$$

Taking the method used for (2.14), we derive the following theorem.

Algorithm 1 Algorithm for Problem (2.8)

- **Step 1**: Solve (2.14) and (2.20) respectively;
- **Step 2**: Compare the results of (2.14) and (2.20), the solution corresponding to the maximum result also solves (2.8)

Theorem 2.2 *The optimal solution for (2.20) is given as*

$$\tau_1^* = \min\{\max\{\frac{bT}{b + P_{th}}, \frac{2cT}{v^* + d - 1}\}, \frac{cT}{c + P_{th}}\}, \tag{2.21}$$

where $v^* > 1$ is the unique solution of $f(v) = d$.

Proof The proof of Theorem 2.2 is similar as the proof of Theorem 2.1 and is ignored here for simplicity.

Based on the conditions that (2.9) and (2.11) are equalities, the optimal consumed energy at the users is given as

$$E_1^* = \eta P h_1 (T - \tau_1^*), \tag{2.22}$$

$$E_2^* = \frac{E_1^* \gamma_1}{\gamma_2} - \frac{P_{th} \tau_1^*}{\gamma_2}. \tag{2.23}$$

Based on the above analysis, the algorithm to solve (2.8) is summarized in Algorithm 1.

2.4 Numerical Results and Discussions

In this section, numerical results are presented to evaluate the performance of WPCNs with NOMA under SIC constraints. This setting of parameters is set as follows. We set the energy harvesting efficiency as $\eta = 0.6$, the noise power at the information receiver as -60 dBm, $T = 1$ s. The available bandwidth is assumed as 1 MHz. Assume the channels are free space and modeled as $h_i = 10^{-3} d_{fi}^{-2}$ and $g_i = 10^{-3} d_{bi}^{-2}$, where d_{fi} is the distance between the power station and U_i, and d_{bi} is the distance between the information receiver and U_i, respectively. The NOMA scheme without SIC constraints in [4] and TDMA scheme in [9] are used as benchmarks. To simplify the description, we denote the NOMA scheme with SIC constraints, the NOMA scheme without SIC constraints and the TDMA scheme as schemes (a), (b) and (c).

To compare the fairness of different schemes, we use the Jain's fairness index, $\mathcal{J}$, which is given by [11]

$$\mathcal{J} = \frac{(\sum_{i=1}^2 R_i)^2}{2 \sum_{i=1}^2 R_i^2}.$$

First, we study the effect of channel gains on the two-user case. For the purpose of exposition, we compare the following three scenarios with different users' locations. In Case 1, referred to the scenario with more distinct channel gains of users ($h_1 = h_2$ and $g_1 > g_2$), $d_{f1} = d_{f2} = 1$ m, $d_{b1} = 2$ m and $d_{b2} = 4$ m. In Case 2, referred to the scenario with less distinct channel gains of users ($h_1 = h_2$ and $g_1 \approx g_2$), $d_{f1} = d_{f2} = 1$ m, $d_{b1} = 2.5$ m, $d_{b2} = 2.6$ m. In Case 3, also referred to the scenario with more distinct channel gains of users ($h_1 < h_2$ and $g_1 > g_2$), $d_{f1} = 2$ m, $d_{f_2} = 1$ m, $d_{b1} = 1$ m, and $d_{b2} = 4$ m.

Figures 2.3, 2.4 and 2.5 show the system rate and the minimum user rate versus the transmit power with different threshold powers. As the transmit power increases, the system rates of the three schemes improve. We first analyze the performances of Figs. 2.3 and 2.5 corresponding to the more distinct channel power gains scenarios. We observe that the performances of Figs. 2.3 and 2.5 are the same, which indicates the system performance depends on the product of h_i and g_i rather than h_i or g_i individually. In addition, the system performance is affected by the threshold power. When the threshold power is low, scheme (a) can achieve the same system performance as scheme (b). When the threshold power increases and the transmit power is low, the system rate of scheme (a) is smaller than that of scheme (b). The explanation is given as follows. For the two-user case via scheme (a), the power control is employed at U_2 to satisfy the SIC constraint. When the threshold power is high and the transmit power is low, the harvested power of U_2 cannot be used up, which degrades the system rate. Moreover, the minimum user rates via schemes (a) and (b) are larger than that via scheme (c). Then, we analyze the performance of Fig. 2.4 corresponding to the less distinct channel gains scenario. Compared with scheme (c), both the system rate and the minimum user rate of scheme (a) are less. It is because keeping the distinctness among users' channel power gains is crucial for NOMA systems to minimize inter-user interference. From Fig. 2.4, we observe that the minimum user rate of scheme (a) may outperform that of scheme (b), especially when the transmit power is not large enough. The explanation is given as follows. For scheme (a), the energy harvesting time is larger than that of scheme (b) since scheme (a) needs to harvest more energy to satisfy the SIC constraints. Hence, U_1 of scheme (a) can transmit data with a relatively large power, which is large enough to compensate the decrease of data transmission time. Compared with scheme (b) where U_1 transmits data with a small power, the minimum user rate of scheme (a) is larger.

Figure 2.6 describes the user fairness versus the transmit power with different threshold powers. From Fig. 2.6a, c, we observe that the NOMA schemes can significantly improve the users' fairness for the more distinct channel gains scenarios. However, for the less distinct channel gains scenario as shown in Fig. 2.6b, the TDMA scheme can achieve the best user fairness. This observation is coincident with Fig. 2.4.

Fig. 2.3 Transmission rate
versus transmit power for
Case 1. (**a**) $P_{\text{th}} = 10\,\text{dB}$. (**b**)
$P_{\text{th}} = 15\,\text{dB}$. (**c**)
$P_{\text{th}} = 20\,\text{dB}$

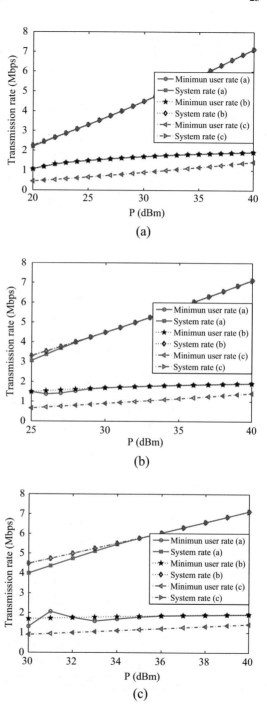

Fig. 2.4 Transmission rate
versus transmit power for
Case 2. (**a**) $P_{th} = 10\,$dB. (**b**)
$P_{th} = 15\,$dB. (**c**)
$P_{th} = 20\,$dB

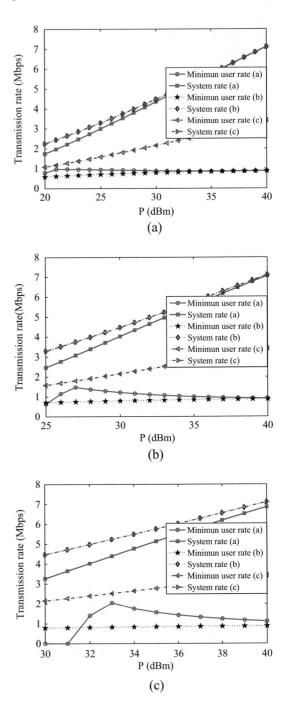

Fig. 2.5 Transmission rate
versus transmit power for
Case 3. (**a**) $P_{th} = 10$ dB. (**b**)
$P_{th} = 15$ dB. (**c**)
$P_{th} = 20$ dB

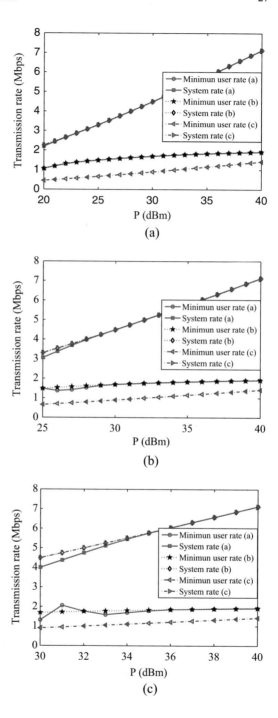

Fig. 2.6 Fairness versus
transmit power. (**a**) Case 1.
(**b**) Case 2. (**c**) Case 3

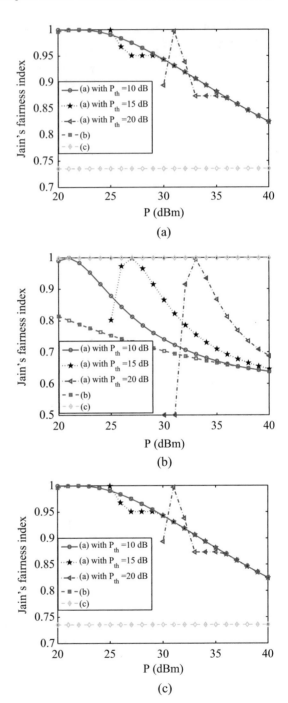

(a)

(b)

(c)

2.5   Conclusions

In this chapter, we has studied a practical wireless powered communication network with NOMA under SIC constraints. We consider a two-users case (Fig. 2.7). To maximize the system throughput, the optimization problem has been formulated to find the optimal tradeoff between WPT and WIT phases, and the optimal energy control at the users. The optimization problem has been divided into two sub-problems, each of which can be solved by exploiting KKT conditions. Then, The optimal solution has been finally derived in a closed-form expression. Simulation results have shown the effect of SIC constraints on the system performance (Fig. 2.8).

Fig. 2.7 Transmission rate versus number of users with $P = 35\,\text{dBm}$. (**a**) $P_{\text{th}} = 10\,\text{dB}$. (**b**) $P_{\text{th}} = 15\,\text{dB}$

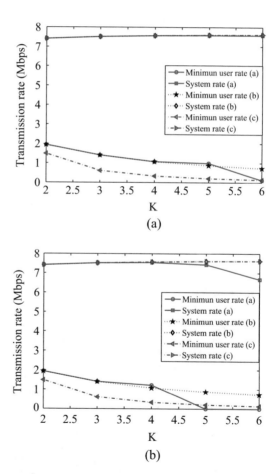

Fig. 2.8 Transmission rate
versus number of users with
$P = 40\,\mathrm{dBm}$. (**a**)
$P_{\mathrm{th}} = 10\,\mathrm{dB}$. (**b**)
$P_{\mathrm{th}} = 15\,\mathrm{dB}$

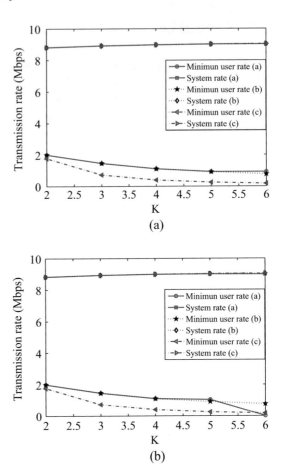

(a)

(b)

References

1. B. Lyu, Z. Yang, and G. Gui, "Non-orthogonal multiple access in wireless powered communication networks with SIC constraints," *IEICE Trans. Commun.* vol. E101-B, no. 4, pp. 1094–1101, Apr. 2018.
2. Y. Saito, A. Benjebbour, Y. Kishiyama, and T. Nakamura, "System-level performance evaluation of downlink non-orthogonal multiple access (NOMA)," in *Proceedings of International Symposium on Personal, Indoor and Mobile Radio Communications*, pp. 611–615, London, UK, 8–11 Sep. 2013.
3. S. M. R. Islam, N. Avazov, O. A. Dobre and K. s. Kwak, "Power-Domain Non-Orthogonal Multiple Access (NOMA) in 5G Systems: Potentials and Challenges," *IEEE Commun. Surveys Tuts.*, vol. 19, no. 2, pp. 721–742, 2nd Quart. 2017.
4. P. D. Diamantoulakis, K. N. Pappi, Z. Ding, and G. K. Karagiannidis, "Wireless powered communications with non-orthogonal multiple access," *IEEE Trans. Wireless Commun.*, vol. 15, no. 15, pp. 8422–8436, Dec. 2016.
5. P. D. Diamantoulakis and G. K. Karagiannidis, "Maximizing proportional fairness in wireless powered communications," *IEEE Wireless Commun. Lett.*, vol. 6, no. 2, pp. 202–205, Apr. 2017.

6. Y. Yuan and Z. Ding, "The application of non-orthogonal multiple access in wireless powered communication networks," in *Proc. SPAWC*, Edinburgh, UK, Aug. 2016.
7. C. Guo, B. Liao, L. Huang, Q. Li, and X. Lin, "Convexity of fairness-aware resource allocation in wireless powered communication networks," *IEEE Commun. Lett.*, vol. 20, no. 3, pp. 474–477, Mar. 2016.
8. M. S. Ali, H. Tabassum, and E. Hossain, "Dynamic user clustering and power allocation for uplink and downlink non-orthogonal multiple access (NOMA) systems," *IEEE Access*, vol. 4, pp. 6325–6343, Oct. 2016.
9. H. Ju and R. Zhang, "Throughput maximization in wireless powered communication networks," *IEEE Trans. Wireless Commun.*, vol. 13, no. 1, pp. 418–428, Jan. 2014.
10. S. Boyd and L. Vandenberghe, *Convex Optimization*. Cambridge University Press, 2004.
11. M. Al-Imari, P. Xiao, M. A. Imran, and R. Tafazolli, "Uplink non-orthogonal multiple access for 5G wireless networks," in *Proc. ISWCS,* pp. 781–785, Barcelona, Spain, 26–29 Aug. 2014.

Chapter 3
Wireless Powered Communication Networks with Backscatter Communication

Abstract By exploiting the advantages of the harvest-then-transmit (HTT) mode and the BackCom mode, two schemes that integrating the HTT mode and the BackCom mode are proposed for the WPCN in this chapter, i.e., the hybrid user scheme and the hybrid mode scheme. For the hybrid user scheme, the WPCN consists of two types of users, which adopt the HTT mode and the BackCom mode, respectively. While, each user can work in either the HTT mode or the BackCom mode for the hybrid mode scheme. The system throughput maximization problems are formulated for the proposed schemes and the optimal solutions are derived, respectively. The optimal solution for the hybrid user scheme shows that only the user in the BackCom mode with the largest backscatter rate can be scheduled. Moreover, the optimal users' working mode permutation is studied for the hybrid mode scheme. Simulation results show the superiority of the proposed schemes over the single mode schemes in terms of system throughput.

3.1 Introduction

At present, WPCNs mostly follow the HTT mode. According to the HTT mode, the users always harvest energy first, and then transmit information using the harvested energy [3]. That is to say, the energy harvesting time is necessary, which must occupy a part of a transmission block. For the users which are far from the HAP, they usually takes a longer time to harvest sufficient energy. When the user has information to transmit but it has not harvested enough energy, the amount of transmitted information will be greatly affected.

As we stated in Chap. 1, the BackCom mode is also a working mode in WPCNs. In the BackCom mode, if the duty cycle [4] is not considered, comparing with the information transmission time, the energy harvesting time is negligible [5]. In other words, the users can reflect the instantaneous incident signal for information transmission. However, the BackCom mode also has its disadvantages. Different

© The Author(s), under exclusive license to Springer Nature Switzerland AG 2019
G. Gui, B. Lyu, *Optimization for Wireless Powered Communication Networks*,
SpringerBriefs in Electrical and Computer Engineering,
https://doi.org/10.1007/978-3-030-01021-8_3

from the HTT mode following which the users transmit information actively, the information backscattering of the users depends on the incident signals. If the incident signals are not available, the users cannot backscatter any information [5, 6].

Considering the characteristics of HTT mode and BackCom mode, this chapter proposes two schemes for WPCNs, i.e., the hybrid user scheme and hybrid mode scheme. For the hybrid user scheme, the WPCN consists of the users working in the BackCom mode and the users working in the HTT mode. It is assumed that all users are scheduled to transmit or backscatter information via TDMA. During the energy harvesting time for the HTT users, all BackCom users can backscatter information. Hence, the entire transmission block can be used for information transmission, which can significantly improve the system throughput. For the hybrid mode scheme, each user can work in either the HTT mode or the BackCom mode. All users also work via TDMA. During each user's allocated time slot, the user sequentially adopts the HTT mode and the BackCom mode for information transmission. Compared with the user only working in one single mode, the flexible permutations of working modes can improve the system performance. For the two proposed schemes, we formulate the optimization problems for the system throughput maximization by finding the optimal time allocation schemes, respectively. For the hybrid user scheme, the closed-form solution is derived, from which we derive that only the BackCom user with the largest backscatter rate can be scheduled. For the hybrid mode scheme, the optimal permutation of working modes is analyzed, from the optimal transmission policy is obtained. Simulation results finally prove that the performance superiority of the proposed schemes. Note that the majority of the contents of this chapter are based on our previous work [1, 2].

3.2 Hybrid User Scheme

3.2.1 System Model

As shown in Fig. 3.1, we consider a WPCN with hybrid user scheme, which includes a two-antenna HAP and K single-antenna nodes. Among these K nodes, the users working in the HTT mode and working in the BackCom mode are respectively denoted as $U_i, i = 1, \cdots, K_h$ and $V_j, j = 1, \cdots, K_b$. The HAP with two antennas works in the FD mode, of which one antenna is used for transferring energy in the downlink and the other antenna is used for receiving signals in the uplink. To make the FD mode work, we assume the perfect SIC techniques [7] are employed at the HAP. Moreover, the HAP is with a stable energy supply, while all users have no initial energy. Denote the downlink channel power gains between the HAP and U_i, and the uplink channel power gains between U_i and the HAP are denoted as h_i and g_i. We assume that h_i and g_i are quasi-static flat fading and remain constant during

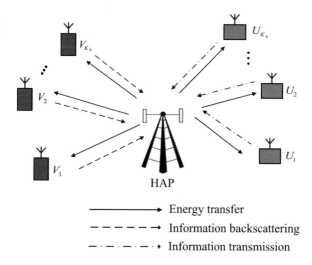

Fig. 3.1 A wireless powered communication network with hybrid user scheme

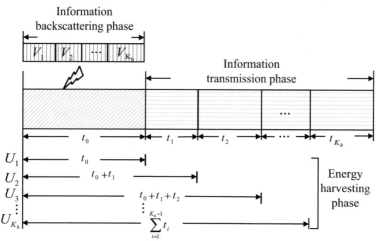

Fig. 3.2 Time block structure for a WPCN with hybrid user scheme

one block but may change from one block to another. It is assumed that the HAP has the prefect information about the channel power gains.

The hybrid user scheme is studied based on time blocks, each of which is given as T seconds. All users backscatter or transmit information via TDMA. The time block is divided into K main time slots, whose durations are denoted as $t_i, i = 0, 1, \cdots, K$, where $t_0 + \sum_{i=1}^{K_h} t_i \leq T$. The first time slot with duration of t_0 is allocated to the BackCom users for information backscattering, which corresponds to the information backscattering phase. The other time slots are allocated to the HTT users, which corresponds to the information transmission phase. The structure of a time block is shown in Fig. 3.2. During the time block, the HAP transits RF signals with P. Due to the high discharge characteristic of supercapacitors, we assume that U_i can only harvest energy before its transmission but not after [9]. Hence, U_i harvests energy during $\sum_{n=0}^{i-1} t_n$ is given by

$$E_i = \eta P h_i \sum_{n=0}^{i-1} t_n, \quad i = 1, \cdots, K_h,$$

where η_i is the energy harvesting efficiency. Based on the harvested energy, U_i transmits information during t_i. We assume that all the harvested energy is used for information transmission during t_i. Hence, the achievable throughput of U_i is expressed as

$$R_i^h = t_i W \log_2 \left(1 + \frac{E_i g_i}{t_i \sigma^2} \right) = t_i W \log_2 \left(1 + \frac{\gamma_i \sum_{n=0}^{i-1} t_n}{t_i} \right), \quad j = 1, \ldots, K_h,$$

$$(3.1)$$

where W is the bandwidth, $\gamma_i = \frac{\eta P h_i g_i}{\sigma^2}$, $i = 1, \cdots, K_h$, σ^2 is the noise power at the HAP.

We assume that all BackCom users backscatter information during t_0 via TDMA. Hence, the first time slot is further divided into K_b time slots, whose durations are given by τ_i and satisfy $\sum_{j=1}^{K_b} \tau_j \leq t_0$. During τ_j, V_j backscatters information to the HAP, while the other BackCom users stay in the idle state. The HAP serves as the coordinator for all users. Denote the backscatter rate of V_j as B_j, which depends on the setting of the RC circuit elements [5] and is given as a constant here. Hence, the achievable throughput of V_j is formulated as

$$R_j^b = \tau_j B_j^b, \, j = 1, \cdots, K_b,$$

$$(3.2)$$

Based on (3.1) and (3.2), the sum-throughput is formulated as

$$R = \sum_{i=1}^{K_h} R_i^h + \sum_{j=1}^{K_b} R_j^b = \sum_{i=1}^{K_h} t_i W \log_2 \left(1 + \frac{\gamma_i \sum_{n=0}^{i-1} t_n}{t_i} \right) + \sum_{j=1}^{K_b} \tau_j B_j^b.$$

$$(3.3)$$

3.2.2 Throughput Maximization

In this section, we study the sum-throughput maximization problem by find the optimal time allocation scheme. Define $t = [t_0, \cdots, t_{K_h}]$ and $\tau = [\tau_1, \cdots, \tau_{K_b}]$. The optimization problem is formulated as

$$\max_{t,\tau} \quad R(t,\tau)$$

$$\text{s.t.} \quad \sum_{i=1}^{K_b} \tau_i \leq t_0,$$

$$t_0 + \sum_{j=1}^{K_h} t_j \leq T, \tag{3.4}$$

$$\tau_i \geq 0, \quad i = 1, \cdots, K_b,$$

$$t_j \geq 0, \quad j = 0, 1, \cdots, K_h.$$

Denote the optimal solution for (3.4) as $t^* = [t_0^*, t_1^*, \cdots, t_{K_h}^*]$ and $\tau^* = [\tau_1^*, \cdots, \tau_{K_b}^*]$. Given t_0, the optimization of the sum-throughput is equivalent to optimizing $\sum_{i=1}^{K_h} R_i^h$ and $\sum_{j=1}^{K_b} R_j^b$ respectively. Hence, with given t_0, we first optimize $\sum_{j=1}^{K_b} R_j^b$ to derive τ^*. To maximize $\sum_{j=1}^{K_b} R_j^b$, we easily derive the following theorem.

Theorem 3.1 *The optimal time allocation for BackCom users satisfies*

$$\tau_j^* = \begin{cases} \tau_0, & j = K \\ 0, & otherwise \end{cases}, \tag{3.5}$$

where $k = \arg\max_j \{B_j\}_{j=1}^{K_b}$.

From Theorem 3.1, it can be seen that only the BackCom user with the maximal backscatter rate can be activated and allocated with t_0, and the other BackCom users cannot work. With Theorem 3.1, R is rewritten as

$$R(t) = \sum_{i=1}^{K_h} t_i W \log_2(1 + \frac{\gamma + i \sum_{n=0}^{i-1} t_n}{t_i}) + t_0 B_k^b, \tag{3.6}$$

and (3.4) is reformulated as

$$\max_{t} \quad R(t)$$

$$\text{s.t.} \quad t_0 + \sum_{j=1}^{K_h} t_j \leq T, \tag{3.7}$$

$$t_j \geq 0, \quad j = 0, 1, \cdots, K_h.$$

It is easy to prove that (3.7) is a convex optimization problem, which can be solved by Lagrange duality. By solving (3.7), its optimal solution is given in Theorem 3.2.

Theorem 3.2

$$t_i^* = \begin{cases} \frac{1}{1+x_i}T, & j = K_h \\ \frac{T - \sum_{n=i+1}^{K_h} t_n^*}{1+x_i}, & i = K_h - 1, \cdots, 1 \\ T - \sum_{n=1}^{K_h} t_n^*, & i = 0 \end{cases},$$

where $x_i = \frac{1}{\gamma_i}(e^{\mathscr{W}(\frac{\gamma_i-1}{e^{c_i+1}})+c_i+1} - 1), i = 1, \cdots, K_t, c_1 = \frac{B_k}{W}ln2, c_i = \frac{B_k}{W}ln2 + \sum_{n=1}^{j-1}\frac{\gamma_n}{\gamma_n x_n+1}, j = 2, \cdots, K_h,$ *and* $\mathscr{W}(.)$ *is the Lambert W-Function* [8].

Proof We use the method given in [9] to solve (3.7). The Lagrangian of (3.7) is given by

$$\mathscr{L}(t, \lambda) = t_0 B_k^b + \sum_{i=1}^{K_h} t_i W \log_2(1 + \frac{\gamma_i \sum_{n=0}^{i-1} t_n}{t_i}) - \lambda(t_0 + \sum_{i=1}^{K_h} t_i - T), \qquad (3.8)$$

where $\lambda \geq 0$ is the Lagrangian multipliers. The dual function of (3.7) is given by

$$\mathscr{G}(\lambda) = \min_{t \in \mathscr{S}} \mathscr{L}(t, \lambda), \qquad (3.9)$$

where $\mathscr{S}$ is the feasible set of t. Since there exists a feasible solution satisfying $t_0 + \sum_{i=1}^{K_h} t_i < T$, the strong duality of (3.7) holds. Hence, (3.7) can be solved by the KKT conditions, which are given by

$$\frac{\partial \mathscr{L}}{\partial t_0} = B_k^b + \sum_{m=1}^{K_h} W \frac{\gamma_m}{ln2(1 + \frac{\gamma_m \sum_{n=0}^{m-1} t_n^*}{t_m^*})} = \lambda^*, \qquad (3.10)$$

$$\frac{\partial \mathscr{L}}{\partial t_i} = W \log_2(1 + \gamma_i \frac{\sum_{n=0}^{i-1} t_n^*}{t_i^*}) - W \frac{\gamma_i \frac{\sum_{n=0}^{i-1} t_n^*}{t_i^*}}{\ln 2(1 + \gamma_i \frac{\sum_{n=0}^{i-1} t_n^*}{t_i^*})}$$

$$+ \sum_{m=i+1}^{K_h} \frac{\gamma_m}{\ln 2(1 + \gamma_m \frac{\sum_{n=0}^{m-1} t_n^*}{t_m^*})} = \lambda^*, i = 1, \cdots, K_h - 1, \qquad (3.11)$$

$$\frac{\partial \mathscr{L}}{\partial t_i} = W \log_2(1 + \gamma_i \frac{\sum_{n=0}^{i-1} t_n^*}{t_i^*}) - W \frac{\gamma_i \frac{\sum_{n=0}^{i-1} t_n^*}{t_i^*}}{\ln 2(1 + \gamma_i \frac{\sum_{n=0}^{i-1} t_n^*}{t_i^*})} = \lambda^*, i = K_h.$$

$$(3.12)$$

Let $\mathscr{B}_i(x_i) = \ln(1 + \gamma_i x_i) - \frac{\gamma_i x_i}{1+\gamma_i x_i}$, $\mathscr{F}_i(x_i) = \mathscr{B}_i(x_i) - \frac{\gamma_i}{1+\gamma_i x_i}$. Substituting (3.10) into (3.11) and (3.12), we have

$$\mathscr{F}_i(x_i) = \mathscr{B}_i(x_i) - \frac{\gamma_i}{1+\gamma_i x_i} - c_i, i = 1, \cdots, K_h, \qquad (3.13)$$

where $c_1 = \frac{B_k}{W} \ln 2$, $c_i = \frac{B_k}{W} \ln 2 + \sum_{m=1}^{i-1} \frac{\gamma_m}{1+\gamma_m x_m}$, $i = 2, \cdots, K_h$. Denote $x_i = \frac{\sum_{n=0}^{i-1} t_n^*}{t_i^*}$. From (3.13), x_i is expressed as

$$x + i = \frac{1}{\gamma_i} \left(e^{W(\frac{\gamma_i - 1}{e^{c_i}+1}) + c_i + 1} - 1 \right), i = 1, \cdots, K_h. \qquad (3.14)$$

From (3.14), we know that x_i is related with c_i. When c_1 is known, it is easy to compute x_1. Based on $c_i = \frac{B_k}{W} \ln 2 + \sum_{m=1}^{i-1} \frac{\gamma_m}{1+\gamma_m x_m}$, $i = 2, \cdots, K_h$, if x_1 is known, we can derive c_2. Similarly, we can derive x_i, $i = 2, \cdots, K_h$. Moreover, we can derive that $\sum_{i=0}^{K_h} t_i^* = T$. Since $x_i = \frac{\sum_{n=0}^{i-1} t_n^*}{t_i^*}$, we can derive the optimal solution as shown in Theorem 3.2. According to [9], the computational complexity of (3.7) is given as $\mathscr{O}(K_n)$.

3.2.3 Numerical Results and Discussions

In this section, simulation results are presented to evaluate the performance of the hybrid user scheme. The simulation environment is set as follows. We assume that there is $K_b = 1$ backscatter user. The bandwidth is set as $W = 1$ MHz. The forward and backward channels are free-space and reciprocal. Let $h_j = g_j = 10^{-3} d_i^{\delta_i} \rho_i^2$, where d_i is the distance between the HAP and U_i, δ_i is the pathloss exponent, ρ_i is the small-term fading, and ρ_i^2 is an exponentially distributed random variable with unit mean. Without loss of generality, let $\delta = \delta_1 = \cdots, \delta_K = 3$ and $T = 1$ s. All of simulation curves are adopted 1000 Monte Carlo runs. For performance comparison, the HHT mode proposed in [9] and the BackCom scheme are used as benchmarks.

Figure 3.3 shows the curves of the average sum-throughput versus the backscatter rate with $K_h = 3$ and $P = 20$ dBm. As the backscatter rate increases, the sum-throughputs achieved by both the BackCom scheme and the hybrid user scheme increase, while the sum-throughput achieved by the HTT scheme does not change. The explanation is given as follows. For the HTT scheme, t_0 is only used for energy harvesting. On the contrary, t_0 is also used for information backscattering for the hybrid user scheme. Moreover, the performance of the hybrid user scheme is also superior to the BackCom scheme. It is because when the BackCom user backscatters information, the HTT users can simultaneously harvest energy for the further information transmission.

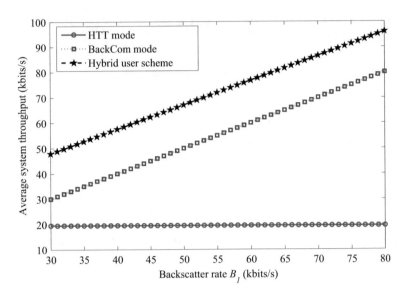

Fig. 3.3 A wireless powered communication network with hybrid user scheme

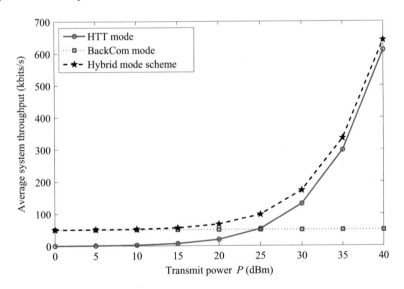

Fig. 3.4 Time block structure for a WPCN with hybrid user scheme

In Fig. 3.4, we investigate the effect of transmit power of the HAP on the average sum-throughput with $B_1 = 50$ kbits/s. It is obvious that the performance of the hybrid user scheme is best. As the transmit power increases, the gap between the HTT scheme and the hybrid user scheme reduces. When the transmit power is small (e.g. smaller than 10 dBm), the sum-throughput achieved by the BackCom scheme is close to that achieved by the hybrid user scheme.

3.3 Hybrid Mode Scheme

3.3.1 System Model

In the section, we consider a WPCN with hybrid mode scheme. The network includes an HAP and K users, where each user can work in either the HTT mode or the BackCom mode. The users are denoted by $U_i, i = 1, \cdots, K$. The channel power gains between the HAP and the users and between the users and the HAP are denoted as h_i and $g_i, i = 1, \cdots, K$. The other settings of the system model are the same as that described in Sect. 3.2.1.

We also study the hybrid mode scheme during a transmission block, the duration of which is T. The block structure for the hybrid mode scheme is described in Fig. 3.5. In the proposed system, the HAP continuously broadcasts energy with constant power, denoted as P, to all users during one block. The block is divided into $K + 1$ main time slots, denoted as $t_i, i = 0, \cdots, K$. U_i is allocated with $t_i, i = 1, \cdots, K$. The main time slot t_i is further divided into three parts, denoted as α_i, β_i and $t_i - \alpha_i - \beta_i, i = 1, \cdots, K$. α_i and β_i are utilized for the HTT mode, and the remaining part is used for the backscatter mode. The structure of a time block is shown in Fig. 3.6. Due to the high discharge characteristic of supercapacitors, we assume that each user can only harvest energy before its transmission but not after [9]. Hence, the harvested energy of U_i, denoted as E_i, over $\sum_{j=0}^{i-1} t_j + a_i$ is given by

$$E_i = \eta_i P h_i (\sum_{j=0}^{i-1} t_j + \alpha_i), \quad i = 1, \cdots, K,$$

where η_i is the efficiency of energy harvesting. Then, U_i transmits information to the HAP during β_i (the data transmission period) with the previously harvested energy.

Fig. 3.5 A wireless powered communication network with hybrid mode scheme

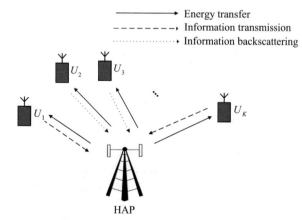

$\longrightarrow$ Energy transfer
$------\rightarrow$ Information transmission
$\cdots\cdots\cdots\rightarrow$ Information backscattering

HAP

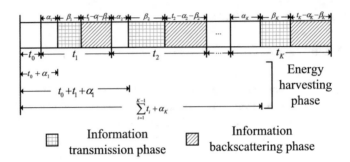

Fig. 3.6 Time block structure for a WPCN with hybrid mode scheme

We also assume that all harvested energy will be exhausted during β_i to avoid energy waste. Hence, the achievable throughput of U_i for the HTT mode is given by

$$R_i^h = \beta_i W \log_2(1 + \frac{E_i g_i}{\beta_i \sigma^2}) = \beta_i W \log_2(1 + \frac{\sum_{j=0}^{i-1} t_j + \alpha_i}{\beta_i} \gamma_i), \quad i = 1, \cdots, K,$$

where W is the bandwidth and $\gamma_i = \frac{\eta_i P h_i g_i}{\sigma^2}$.

Denote the backscatter rate of U_i as B_i^b. The throughput of U_i for the BackCom mode over $t_i - \alpha_i - \beta_i$ is given by

$$R_i^b = (t_i - \alpha_i - \beta_i) B_i^b, \quad i = 1, \cdots, K.$$

Based on the above analysis, the sum-throughput of all users is given by

$$R = \sum_{i=1}^{K} R_i^h + \sum_{i=1}^{K} R_i^b.$$

3.3.2   Throughput Maximization

In this section, we study the sum-throughput maximization of the hybrid mode scheme by optimizing the transmission policy. An optimization problem is first formulated. However, it is not easy to solve the optimization problem with multiple users, particularly when the number of users exceeds two. Hence, we first consider a special case, i.e., $K = 2$, to obtain some insights. Based on these insights, the general case ($K \geq 3$) is further solved. Denote $t = [t_0, \cdots, t_K]$, $\alpha = [\alpha_1, \cdots, \alpha_K]$, and $\beta = [\beta_1, \cdots, \beta_K]$. The optimization problem is formulated as

$$\max_{t,\alpha,\beta} \quad R(t,\alpha,\beta)$$

$$\text{s.t.} \quad \sum_{i=0}^{K} t_i \leq T,$$

$$\alpha_i + \beta_i \leq t_i, \qquad i = 1, \cdots, K,$$
$$\alpha_i \geq 0, \qquad i = 1, \cdots, K, \qquad\qquad (3.15)$$
$$\beta_i \geq 0, \qquad i = 1, \cdots, K,$$
$$t_i \geq 0, \qquad i = 0, 1, \cdots, K.$$

It can be proved that (3.15) is an optimization problem. Denote the optimal solution for (3.15) as $t^* = [t_0^*, \cdots, t_K^*]$, $\alpha^* = [\alpha_1^*, \cdots, \alpha_K^*]$, and $\beta^* = [\beta_1^*, \cdots, \beta_K^*]$. First, a useful structure of the optimal solution for (3.15) is given in the following lemma.

Lemma 3.1 *The optimal solution for (3.15) satisfies*

$$\sum_{i=0}^{K} t_i^* = T, \qquad\qquad (3.16)$$

$$\alpha_i^* = 0, \ i = 1, \cdots, K. \qquad\qquad (3.17)$$

Proof Lemma 3.1 is proven by contradiction. Define the optimal solution for Problem P1 as $\bar{t} = [\bar{t}_1, \cdots, \bar{t}_K]$, $\bar{\alpha} = [\bar{\alpha}_1, \cdots, \bar{\alpha}_K]$, and $\bar{\beta} = [\bar{\beta}_1, \cdots, \bar{\beta}_K]$. The proof of (3.16) is similar as the proof Lemma 2.2 and is omitted for simplicity. Then, we prove that $\bar{\alpha}$ with $\alpha_k \neq 0, k \in [1, \cdots, K]$ is not the optimal solution. Consider t^*, α^* and β^*, which satisfy the following conditions: $\alpha_i^* = 0, \ i = 1, \cdots, K, \ t_{k-1}^* = \bar{t}_{k-1} + \bar{\alpha}_k$ and $t_k^* = \bar{t}_k - \bar{\alpha}_k, \ t_i^* = \bar{t}_i, i \neq k - 1, k$, and $\beta^* = \bar{\beta}$. For $k = 1$, the harvested energy and allocated time for transmitting or backscattering information of $U_i, i = 1, \cdots, K$ with t^*, α^*, and β^* are the same as those of U_i with $\bar{t}, \bar{\alpha}$, and $\bar{\beta}$. Hence, t^*, α^*, and β^* is also one optimal solution. For $k \geq 2$, $R_{k-1}^h(t^*,\alpha^*,\beta^*) = R_{k-1}^h(\bar{t},\bar{\alpha},\bar{\beta})$, $R_{k-1}^b(t^*,\alpha^*,\beta^*) = (t_{k-1}^* - \beta_{k-1}^*)B_{k-1}^b = (\bar{t}_{k-1} + \bar{\alpha}_k - \bar{\beta}_{k-1})B_{k-1}^b > (\bar{t}_{k-1} - \bar{\beta}_{k-1})B_{k-1}^b = R_{k-1}^b(\bar{t},\bar{\alpha},\bar{\beta})$; $R_k^h(t^*,\alpha^*,\beta^*) = R_k^h(\bar{t},\bar{\alpha},\bar{\beta})$, $R_k^b(t^*,\alpha^*,\beta^*) = (t_k^* - \beta_k^*)B_k^b = (\bar{t}_k - \bar{\alpha}_k - \bar{\beta}_k)B_k^b = R_k^b(\bar{t},\bar{\alpha},\bar{\beta})$; for $i \neq k - 1, k$, $R_i^h(t^*,\alpha^*,\beta^*) = R_i^h(\bar{t},\bar{\alpha},\bar{\beta})$ and $R_i^b(t^*,\alpha^*,\beta^*) = R_i^b(\bar{t},\bar{\alpha},\bar{\beta})$. Hence, we derive that $R(t^*,\alpha^*,\beta^*) > R(\bar{t},\bar{\alpha},\bar{\beta})$, which contradicts the assumption that $\bar{\alpha}$ is the optimal solution. This completes the proof.

With $\alpha_i^* = 0, i = 1, \cdots, K$, the sum-throughput is re-expressed as

$$R(t,\beta) = \sum_{i=1}^{K} \beta_i W \log_2\left(1 + \frac{\sum_{j=0}^{i-1} t_j}{\beta_i}\gamma_i\right) + \sum_{i=1}^{K}(t_i - \beta_i)B_i^b.$$

Equation (3.15) can be reduced to the following problem

$$\max_{t,\beta} \quad R(t, \beta)$$

$$\text{s.t.} \quad \sum_{i=0}^{K} t_i = T,$$

$$\beta_i \leq t_i, \quad i = 1, \cdots, K,$$

$$\beta_i \geq 0, \quad i = 1, \cdots, K,$$

$$t_i \geq 0, \quad i = 0, 1, \cdots, K.$$

(3.18)

Before solving (3.18) directly, a two-user case is first considered.

3.3.3   Two-User Case

To solve (3.18) with $K = 2$, we adopt a two-stage approach. Let $\bar{T} = t_0 + t_1$. From [11], we know that each user can work in either HTT mode or backscatter mode during the given slot. First, given $\bar{T}$, we study the optimal working mode of U_1, corresponding to the single-user case [11]. Then, with the derived result of U_1, we further study the working mode of U_2. To facilitate the understanding of this paper, we still present the details of solving the single-user case.

3.3.3.1   The Optimal Solution for Single-User Case

With given $\bar{T}$, the optimization problem of the single-user case is given as follows.

$$\max_{t_0,t_1,\beta_1} \quad \beta_1 W \log_2(1 + \frac{t_0}{\beta_1}\gamma_1) + (t_1 - \beta_1)B_1^b$$

$$\text{s.t.} \quad t_0 + t_1 = \bar{T},$$

$$0 \leq \beta_1 \leq t_1,$$

$$t_0 \geq 0.$$

(3.19)

To solve (3.19), we first derive the relationship between t_0 and β_1. Let $\bar{\bar{T}} = t_0 + \beta_1$. The relationship between t_0 and β_1 can be derived by solving the following problem

$$\max_{t_0,\beta_1} \quad \beta_1 W \log_2(1 + \frac{t_0}{\beta_1}\gamma_1)$$

$$\text{s.t.} \quad t_0 + \beta_1 = \bar{\bar{T}},$$

$$t_0, \beta_1 \geq 0.$$

(3.20)

The solution for (3.20) can be obtained using the method presented in [3]. With the given $\bar{\bar{T}}$, the optimal solution for (3.20) is given as

$$t_0 = \frac{z_1^* - 1}{\gamma_1 + z_1^* - 1} \bar{\bar{T}} \text{ and } \beta_1 = \frac{\gamma_1}{\gamma_1 + z_1^* - 1} \bar{\bar{T}},$$

where $z_1^* > 1$ is the corresponding solution of $f_1(z_1) = \gamma_1$ and $f_1(z_1)$ is defined as $f_1(z_1) = z_1 \ln z_1 - z_1 + 1$. Let $B_1^t = \frac{\gamma_1}{\gamma_1 + z_1^* - 1} W \log_2 z_1^*$, which is the normalized throughput of U_1 with the HTT mode. With the above result, (3.19) is reduced to (3.21), which is given by

$$\max_{\bar{\bar{T}}} \quad B_1^t \bar{\bar{T}} + (\bar{T} - \bar{\bar{T}}) B_1^b$$

$$\text{s.t.} \quad 0 \le \bar{\bar{T}} \le \bar{T}. \tag{3.21}$$

The solution for (3.21) straightforward to derive and is given as follows.

$$\bar{\bar{T}} = \begin{cases} \bar{T}, & B_1^t \ge B_1^b, \\ 0, & B_1^t < B_1^b. \end{cases} \tag{3.22}$$

If $\bar{\bar{T}} = \bar{T}$, then U_1 works in the HTT mode; otherwise, U_1 works in the backscatter mode.

3.3.3.2 The Optimal Solution for Two-User Case

With the working mode of U_1, (3.18) is further studied to find the optimal working mode of U_2. Consider two sub-cases, i.e., U_1 works in the HHT mode or the backscatter mode. If U_1 works in the HTT mode, then the sum-throughput is rewritten as $R(\bar{T}, t_2, \beta_2) = B_1^t \bar{T} + \beta_2 W \log_2(1 + \frac{\bar{T}}{\beta_2} \gamma_2) + (t_2 - \beta_2) B_2^b$. With the updated $R(\bar{T}, t_2, \beta_2)$, (3.18) is rewritten as

$$\max_{\bar{T}, t_2, \beta_2} \quad R(\bar{T}, t_2, \beta_2)$$

$$\text{s.t.} \quad \bar{T} + t_2 = T,$$

$$0 \le \beta_2 \le t_2, \tag{3.23}$$

$$\bar{T} \ge 0.$$

To solve (3.23), a method similar to that used for (3.19) is adopted. Let $\hat{T} = \bar{T} + \beta_2$. First, we determine the relationship between $\bar{T}$ and β_2 by solving the following problem

$$\max_{\bar{T}, \beta_2} \quad B_1^t \bar{T} + \beta_2 W \log_2(1 + \frac{\bar{T}}{\beta_2} \gamma_2)$$

$$\text{s.t.} \quad \bar{T} + \beta_2 = \hat{T},$$

$$\bar{T}, \beta_2 \geq 0. \tag{3.24}$$

Using a method similar to that for (3.20), we derive the optimal solution for (3.24) with the given $\hat{T}$, which is as follows

$$\bar{T} = \frac{z_2^* - 1}{\gamma_2 + z_2^* - 1} \hat{T} \quad \text{and} \quad \beta_2 = \frac{\gamma_2}{\gamma_2 + z_2^* - 1} \hat{T},$$

where z_2^* is the corresponding solution of $f_2(z_2) = \gamma_2$ and $f_2(z_2)$ is defined as $f_2(z_2) = z_2 \ln z_2 - (1 + \frac{\ln 2}{W} B_1^t) z_2 + 1$. Let $B_2^t = \frac{z_2^* - 1}{\gamma_2 + z_2^* - 1} B_1^t + \frac{\gamma_2}{\gamma_2 + z_2^* - 1} W \log_2 z_2^*$. Hence, (3.23) is reduced to (3.25), which is formulated as

$$\max_{\hat{T}} \quad B_2^t \hat{T} + (T - \hat{T}) B_2^b$$

$$\text{s.t.} \quad 0 \leq \hat{T} \leq T. \tag{3.25}$$

The result for (3.25) is easily derived and is given as follows.

$$\hat{T} = \begin{cases} T, & B_2^t \geq B_2^b, \\ 0, & B_2^t < B_2^b. \end{cases} \tag{3.26}$$

Combining the results from (3.22) and (3.26), the permutations of two users' working mode are summarized as follows. If $B_1^t \geq B_1^b$ and $B_2^t \geq B_2^b$, then both U_1 and U_2 work in the HTT mode. If $B_1^t \geq B_1^b$ and $B_2^t < B_2^b$, then U_1 cannot work and U_2 works in the backscatter mode with the entire block. To simplify the description, we define the user cannot work, the user works in the backscatter mode and the user works in the HTT mode as "0", "B" and "H", respectively. Therefore, the permutations of this sub-case are given by "HH" and "0B".

For symbol simplification, the symbols defined in the previous sub-case are still used in this sub-case corresponding to U_1 works in the backscatter mode. Since U_1 works in the backscatter mode, R is rewritten as $R(\bar{T}, t_2, \beta_2) = B_1^b \bar{T} + \beta_2 W \log_2(1 + \frac{\bar{T}}{\beta_2} \gamma_2) + (t_2 - \beta_2) B_2^b$. In this sub-case, with the updated $R(\bar{T}, t_2, \beta_2)$, we also solve Problem P6 to derive the optimal solution. Hence, $\bar{T}$, β_2 and $\hat{T}$ in this sub-case are rewritten as

$$\bar{T} = \frac{z_3^* - 1}{\gamma_2 + z_3^* - 1} \hat{T} \quad \text{and} \quad \beta_2 = \frac{\gamma_2}{\gamma_2 + z_3^* - 1} \hat{T},$$

$$\hat{T} = \begin{cases} T, & \bar{B}_2^t \geq B_2^b, \\ 0, & \bar{B}_2^t < B_2^b, \end{cases} \tag{3.27}$$

where z_3^* is the corresponding solution of $f_3(z_3) = \gamma_2$ and $f_3(z_3)$ is defined as
$f_3(z_3) = z_3 \ln z_3 - (1 + \frac{\ln 2}{W} B_1^b)z_3 + 1$, $\bar{B}_2^t = \frac{z_3^*-1}{\gamma_2+z_3^*-1} B_1^b + \frac{\gamma_2}{\gamma_2+z_3^*-1} W \log_2 z_3^*$.

Consider both (3.22) and (3.27). If $B_1^t < B_1^b$ and $B_2^t \geq B_2^b$, then U_1 works in the backscatter mode and U_2 works in the HTT mode. If $B_1^t < B_1^b$ and $B_2^t < B_2^b$, then only U_2 works and stays in the backscatter mode. The permutations of this sub-case are summarized as "BH" and "0B".

Considering the above two sub-cases, the optimal time allocation policy for the two-user case is given by the following proposition.

Theorem 3.3 *For the two-user case, one optimal solution for* (3.18) *is given by*

$$t_0 = \begin{cases} 0, & \text{if "0B" or "BH"}, \\ \frac{z_1^*-1}{\gamma_1+z_1^*-1} \frac{z_2^*-1}{\gamma_2+z_2^*-1} T, & \text{if "HH"}, \end{cases}$$

$$\beta_1 = \begin{cases} 0, & \text{if "0B" or "BH"}, \\ \frac{\gamma_1}{\gamma_1+z_1^*-1} \frac{z_2^*-1}{\gamma_2+z_2^*-1} T, & \text{if "HH"}, \end{cases}$$

$$t_1 = \begin{cases} 0, & \text{if "0B"}, \\ \frac{z_3^*-1}{\gamma_2+z_3^*-1} T, & \text{if "BH"}, \\ \frac{\gamma_1}{\gamma_1+z_1^*-1} \frac{z_2^*-1}{\gamma_2+z_2^*-1} T, & \text{if "HH"}, \end{cases}$$

$$\beta_2 = \begin{cases} 0, & \text{if "0B"}, \\ \frac{\gamma_2}{\gamma_2+z_3^*-1} T, & \text{if "BH"}, \\ \frac{\gamma_2}{\gamma_2+z_2^*-1} T & \text{if "HH"}, \end{cases}$$

$$t_2 = \begin{cases} T, & \text{if "0B"} \\ \frac{\gamma_2}{\gamma_2+z_3^*-1} T, & \text{if "BH"}, \\ \frac{\gamma_2}{\gamma_2+z_2^*-1} T, & \text{if "HH"}. \end{cases}$$

Note that Proposition 3.3 only provides one of the optimal transmission policies for the two-user case. This is because when one of the special conditions, i.e., $B_1^t = B_1^b$, $B_2^t = B_2^b$ and $\bar{B}_2^t = B_2^b$ is satisfied, we can also find other solutions to achieve the maximum sum-throughput. However, the result given in Theorem 3.3 can provide us with some interesting insights. Hereinafter, without specific description, we do not consider the special conditions.

In summary, for a BAWPCN with two users, there are three possible optimal permutations of the working mode, i.e., "0B", "BH" and "HH".

With Theorem 3.3 and the above analysis, we can derive the following corollaries.

Corollary 3.1 *In the optimal condition, the last user could be scheduled and work in only one mode.*

For simplification, we briefly discuss the proof of Corollary 3.1 in the following remark.

Remark 3.1 For the two-user case, it has been proved that the last user could be scheduled and work in only one mode in the optimal condition. Considering the three-user case, with the given working modes of the first two users, it can be seen that the time allocation between the HTT and backscatter modes of the third user is a linear problem, which is similar as Problem P8. To solve the problem, we obtain the conclusion that the third user could be scheduled and work in only one mode in the optimal condition. Similarly, for K-user case ($K \geq 4$), the same conclusion can be obtained.

Corollary 3.2 *In the optimal condition, if the last user works in the backscatter mode, one optimal solution is to allocate the entire block time to U_K, i.e., the first $K - 1$ users are not scheduled.*

Proof Define the optimal normalized throughput for the first $K - 1$ users as B_{K-1}^t. Additionally, define z_K^* as the optimal solution for $f_K(z_K) = \gamma_K$ and $f_K(z_K) = z_K \ln z_K - (1 + \frac{\ln 2}{W} B_{K-1}^t) z_K + 1$. We derive that if U_K can work in the backscatter mode in the optimal condition, then the condition $\frac{z_K^* - 1}{\gamma_K + z_K^* - 1} B_{K-1}^t + \frac{\gamma_K}{\gamma_K + z_K^* - 1} W \log_2 z_K^* \leq B_K^b$ is satisfied. Hence, in the optimal condition, we can let the total time allocated to the first $K - 1$ users be zero, leading to the desired result.

From the above result, we observe that the schedule order of users may affect the time allocation policy. To explain this observation, the following example is given. For "0B" corresponding to the condition $B_1^t \geq B_1^b$ and $B_2^t < B_2^b$, it can achieve a larger sum-throughput with a reversed schedule order, i.e., U_2 is first scheduled. Next, we analyze the sum-throughput of the reversed schedule order scheme. According to [9], as the number of users increases, the normalized sum-throughput increases. Since $B_2^b > B_2^t$, we can derive that $B_2^b > \frac{\gamma_2}{\gamma_2 + z_1^\dagger - 1} W \log_2 z_1^\dagger$, where $\frac{\gamma_2}{\gamma_2 + z_1^\dagger - 1} W \log_2 z_1^\dagger$ is the normalized throughput of U_2 with the HTT mode, and $z_1^\dagger$ is the solution of $f_1(z_1) = \gamma_2$. Additionally, with $B_2^b > B_1^b$, we can derive that the optimal permutation of the reversed schedule order scheme is given as "BH". From [2], the optimal solution for the reversed schedule order scheme indeed provides a larger sum-throughput.

3.3.4 Multi-user Case ($K \geq 3$)

In this sub-section, we propose an algorithm with lower complexity to solve (3.18) if $K \geq 3$. With the result for the two-user case, we can derive all possible optimal permutations of the working mode, one of which is the globally optimal permutation. For the two-user case, there are three permutations, i.e., "0B", "BH" and "HH". When we consider the three-user case, from Corollary 3.1, there are six possible optimal permutations of the working mode, which are given as follows: "0BB", "0BH", "BHB", "BHH", "HHB" and "HHH". From Corollary 3.2, the permutations "0BB", "BHB" and "HHB" can be reduced to the permutation "00B" without affecting the optimal solution. The updated permutations are summarized as "00B", "0BH", "BHH", and "HHH". Similarly, for the four-user case, the possible optimal permutations are "000B", "00BH", "0BHH", "BHHH", and "HHHH". Hence, we can conclude that the K-user case finally has $K + 1$ possible optimal permutations. Since each permutation corresponds to a sub-problem, (3.18) can be decomposed into $K + 1$ sub-problems. By solving these sub-problems sequentially, the solution for the sub-problem with the maximum sum-throughput also solves (3.18).

The analysis for the $K + 1$ sub-problems is given as follows. The $K + 1$ sub-problems can be classified into three types. The first type is that only the last user works and stays in the backscatter mode. The second one is that the first working user adopts the backscatter mode and the other working users adopt the HTT mode. The third one is that all users work in the HTT mode. For the first type of sub-problem, it is easy to derive that the entire block time will be allocated to U_K. For the second type of sub-problems, assume that the number of working users is M. Hence, the index of the first working user, which is in the backscatter mode, is $K - M + 1$. Let $N = K - M + 1$. Since the permutation of the working mode is given, to maximize the sum-throughput, it is straightforward to derive that $t_i = 0, i = 0, 1, , \cdots, K - M$, $\beta_i = 0, i = 1, \cdots, N$ and $\beta_i = t_i, i = N + 1, \cdots, K$. Hence, (3.18) is rewritten as

$$\max_{\{t_i\}_{i=N}^{K}} \quad t_N B_N^b + \sum_{i=N+1}^{K} t_i W \log_2(1 + \frac{\sum_{j=N}^{i-1} t_j}{t_i} \gamma_i)$$

$$\text{s.t.} \quad \sum_{i=N}^{K} t_i = T, \tag{3.28}$$

$$t_i \geq 0, \quad i = N, N+1, \cdots, K.$$

Using convex optimization techniques [10], the optimal time allocation policy for the second type of sub-problems is given in the following theorem.

Theorem 3.4 *The optimal solution for (3.28) satisfies*

$$
t_i^* = \begin{cases}
\frac{1}{1+v_i} T, & i = K, \\
\frac{T - \sum_{n=i+1}^{K} t_n^*}{1+v_i}, & i = K-1, \cdots, N+1, \\
T - \sum_{n=N+1}^{K} t_n^*, & i = N,
\end{cases}
$$

where $v_i = \frac{1}{\gamma_i}(e^{\mathscr{W}(\frac{\gamma_i-1}{e^{a_i}+1})+a_i+1} - 1), i = N+1, \cdots, K, a_{N+1} = \frac{B_N^b}{W} \ln 2, a_i = \frac{B_N^b}{W} \ln 2 + \sum_{n=N+1}^{i-1} \frac{\gamma_n}{\gamma_n v_n+1}, i = N+2, \cdots, K,$ *and* $\mathscr{W}(.)$ *is the Lambert W-Function* [8].

Proof Please refer to the proof of Theorem 3.2.

For the third type of sub-problem, since all users work in the HTT mode, $\beta_i = t_i, i = 1, \cdots, K$. Equation (3.18) is reduced to (3.29), which was studied in [9].

$$
\max_{t} \quad \sum_{i=1}^{K} t_i W \log_2\left(1 + \frac{\sum_{j=0}^{i-1} t_j}{t_i} \gamma_i\right)
$$

$$
\text{s.t.} \quad \sum_{i=0}^{K} t_i = T, \tag{3.29}
$$

$$
t_i \geq 0, \quad i = 0, 1, \cdots, K.
$$

Using the algorithm proposed in [9], the optimal time allocation policy for this type of sub-problem is given in the following theorem.

Theorem 3.5 *The optimal solution for (3.29) satisfies*

$$
t_i^* = \begin{cases}
\frac{1}{1+\mu_i} T, & i = K, \\
\frac{T - \sum_{n=i+1}^{K} t_n^*}{1+\mu_i}, & i = K-1, \cdots, 1, \\
T - \sum_{n=1}^{K} t_n^*, & j = 0,
\end{cases}
$$

where $\mu_i = \frac{1}{\gamma_i}(e^{\mathscr{W}(\frac{\gamma_i-1}{e^{c_i}+1})+c_i+1} - 1), i = 1, \cdots, K, c_1 = 0,$ *and* $c_i = \sum_{n=1}^{i-1} \frac{\gamma_n}{\gamma_n \mu_n+1}, i = 2, \cdots, K.$

Algorithm 2 Algorithm for (3.18)

- **Step 1**: Determine $K + 1$ possible optimal permutations. Based on these permutations, (3.18) is decomposed into $K + 1$ sub-problems;
- **Step 2**: Solve these sub-problems sequentially;
- **Step 3**: Comparing the sum-throughputs of $K + 1$ sub-problems, the solution corresponding to the maximum sum-throughput is also the solution for (3.18).

After solving $K + 1$ sub-problems, the solution corresponding to the maximum sum-throughput also solves (3.18).

The algorithm to solve (3.18) is summarized in Algorithm 2.

Remark 3.2 For the special case in which the backscatter rate of each user is the same, i.e., $B_1 = \cdots = B_K$, the computational complexity of solving Problem P2 can be further reduced. The explanation is given by taking the three-user case as an example. For the three-user case, all possible optimal permutations are given as "00B", "0BH", "BHH", and "HHH". From [1], with the increase of the number of users, the sum-throughput of the second type of sub-problems increases. Moreover, "00B" can be viewed as a special second type of sub-problem. Hence, the sum-throughput of "BHH" is larger than that of "00B" or "0BH". In other words, we only need to solve two sub-problems corresponding to "BHH" and "HHH", respectively. Extending to the case with K ($K \geq 3$) users, we only need to consider the two sub-problems in which all users are scheduled.

3.3.5 Numerical Results and Discussions

In this section, we given the numerical results to evaluate the performance of the hybrid mode scheme. The parameter settings are the same as Sect. 3.2.3.

Figure 3.7 investigates the effect of schedule order for the two-user case. Let $\kappa = 3$, $d_1 = 5$ m, $d_2 = 10$ m, $B_1^b = 20$ kbits/s, and $B_2^b = 60$ kbits/s. It can be observed that the performance of the reversed order scheme is superior to that of the given order scheme, which indicates that the schedule order indeed affects the sum-throughput. Moreover, as the transmit power increases, the sum-throughput of both schedule order schemes increases.

Figure 3.8 shows the average probability of possible optimal permutations versus the transmit power for the three-user case with $\kappa = 3$ and $d_i = 5 + \frac{D_K}{K} * i$, where $D_k = 5$ m. Three scenarios in which users have different backscatter rates are

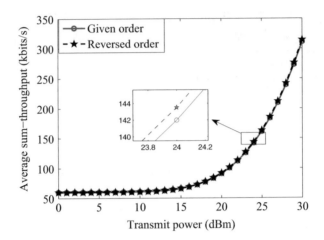

Fig. 3.7 Average sum-throughput vs. transmit power for two-user case with $\kappa = 3$, $d_1 = 5$ m, $d_2 = 10$ m, $B_1^b = 20$ kbits/s, and $B_2^b = 60$ kbits/s

considered. Let $\boldsymbol{B}^b = [B_1^b, B_2^b, B_3^b]$. The three scenarios are given as follows: $\boldsymbol{B}^b = [40, 40, 40]$ kbits/s, $\boldsymbol{B}^b = [20, 40, 60]$ kbits/s, and $\boldsymbol{B}^b = [60, 40, 20]$ kbits/s, which correspond to Fig. 3.8a, b, and c, respectively. Figure 3.8a, c indicate that when the transmit power is low, "BHH" is the optimal permutation. This result occurs because with a low transmit power, the normalized throughput of U_1 adopting the HHT mode is lower than its backscatter rate. Conversely, the optimal permutation of Fig. 3.8b with a low transmit power is "00B" since B_3^b is quite large. For the three scenarios, as the transmit power increases, "HHH" tends to be the optimal permutation with a high probability.

3.4 Conclusions

In this chapter, we have proposed a hybrid user scheme and a hybrid mode scheme for WPCN. We have given the system model for each scheme, where the HAP adopts the FH mode and the users satisfy the energy causality constraints. For the hybrid user scheme, the BackCom users backscatter information during the energy harvesting time of the HTT users, which thus make full use of time for information transmission. For the hybrid mode scheme, each user adopt the HTT mode to transmit information or the BackCom mode to backscatter information flexibly. The system throughput maximization problems have been studied for both schemes and the optimal time allocations in closed form have been derived. Simulation results haven shown the superiority of the proposed two schemes.

Fig. 3.8 Average ratio of optimal permutation vs. transmit power for three-user case with $\kappa = 3$, $d_i = 5 + \frac{D_K}{K} * i$ and $D_k = 5$ m. (**a**) Scenario one with $B_1^b = 40$ kbits/s, $B_2^b = 40$ kbits/s and $B_3^b = 40$ kbits/s. (**b**) Scenario two with $B_1^b = 20$ kbits/s, $B_2^b = 40$ kbits/s and $B_3^b = 60$ kbits/s. (**c**) Scenario three with $B_1^b = 60$ kbits/s, $B_2^b = 40$ kbits/s and $B_3^b = 20$ kbits/s

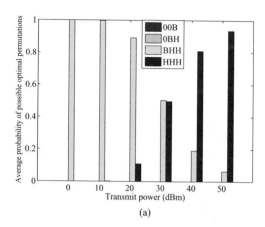

(a)

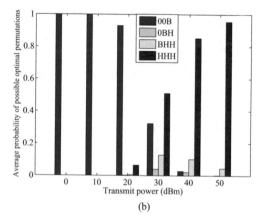

(b)

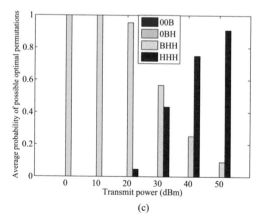

(c)

References

1. B. Lyu, Z. Yang, G. Gui, and Y. Feng, "Throughput maximization in backscatter assisted wireless powered communication networks," *IEICE Trans. Fundamentals*, vol. E100-A, no. 6, pp. 1–5, Jun. 2017.
2. B. Lyu, Z. Yang, G. Gui, and Y. Feng, "Wireless powered communication networks assisted by backscatter communication," *IEEE Access*, vol. 5, pp. 7254–7262, 2017.
3. H. Ju and R. Zhang, "Throughput maximization in wireless powered communication networks," *IEEE Trans. Wireless Commun.*, vol. 13, no. 1, pp. 418–428, Jan. 2014.
4. C. Boyer and S. Roy, "Backscatter communication and RFID: Coding, energy, and MIMO analysis," *IEEE Trans. Commun.*, vol. 62, no. 3, pp. 770–785, Mar. 2014.
5. V. Liu, A. Parks, V. Talla, S. Gollakota, D. Wetherall, and J. R. Smith, "Ambient backscatter: Wireless communication out of thin air," in *Proc. ACM SIGCOMM*, pp. 39–50, Hong Kong, Aug. 2013.
6. J. Kimionis, A. Bletsas, and J. N. Sahalos, "Increased range bistatic scatter radio," *IEEE Trans. Commun.*, vol. 62, no. 3, pp. 1091–1104, Mar. 2014.
7. D. Bharadia, E.McMilin, and S. Katti, "Full duplex radios," in *Proc. ACM SIGCOMM*, pp. 375–386, Hong Kong, Aug. 2013.
8. "Lambert W-Function." [Online]. Available: http://mathworld.wolfram.com/LambertW-Function.html
9. X. Kang, C. K. Ho, and S. Sun, "Full-duplex wireless-powered communication network with energy causality," *IEEE Trans. Wireless Commun.*, vol. 14, no. 10, pp. 5539–5551, Oct. 2015.
10. S. Boyd and L. Vandenberghe, *Convex Optimization*. Cambridge University Press, 2004.
11. D. T. Hoang, D. Niyato, P. Wang, D. I. Kim, and Z. Han, "Ambient backscatter: A new approach to improve network performance for RF-powered cognitive radio networks," *IEEE Trans. Commun.*, vol. 65, no. 9, pp. 3659–3674, Sept. 2017.

Chapter 4
Cognitive Wireless Powered Communication Networks with Hybrid Backscatter Communication

Abstract In this chapter, a hybrid mode scheme is proposed for a CWPCN. The CWPCN consists of a primary communication system and a secondary communication system, where the cognitive user (CU) can adopt the HTT mode, the AB mode or the BB mode following the proposed hybrid mode scheme. The primary transmitter (PT) and the PB serve as the incident signal sources of the AB mode and the BB mode for the CU's information backscattering, respectively. When the primary channel is idle, the CU use the harvested energy from the PT and PB to transmit information to the IR following the HTT mode. The optimal time allocation among the three modes is investigated for the sake of maximizing the throughput of the secondary communication system. The closed-form solution is derived and the optimal combination of the working modes is obtained. Numerical results demonstrate the advantage of the proposed scheme over the single mode schemes in terms of system throughput.

4.1 Introduction

With the rapid development of wireless communication, spectrum resources are becoming increasingly scarce. In the future, WPCN will be widely applied in IoT. It is not possible to allocate dedicated spectrum for each WPCN. Hence, the combined application of WPCN and CR has attracted wide attention of researchers. In [2], a WPCN with cognitive function, termed CWPCN, was first proposed. In this proposed model, the WPCN is the secondary communication system, which shares the spectrum with the primary communication system. In [3], the authors studied the overlay-CWPCN, where the CU can works either in the HTT mode or the AB mode. Note that the AB mode is suitable for the scenario that the CU are far from the RF source (e.g., the PT). When the primary channel is busy, the PT is the incident signal source for the CU and the CU thus works in the AB mode. If there exists a PB around the CU, the PB can also be the incident signal source for the BackCom mode. Since the CU are typically placed nearer to the PB than the PT, the CU thus can be activated to adopt the BB mode to backscatter information [4].

© The Author(s), under exclusive license to Springer Nature Switzerland AG 2019
G. Gui, B. Lyu, *Optimization for Wireless Powered Communication Networks*,
SpringerBriefs in Electrical and Computer Engineering,
https://doi.org/10.1007/978-3-030-01021-8_4

In this chapter, we consider a interweave-CWPCN defined in [2], which consists of a PT and a PB. In this network, we assume that the working states of the primary channel can be first detected via some advanced spectrum sensing techniques. When the PT transmits signals, the PB stays in the idle state, whereas the PB transmit energy signals. To fully exploit the PT and the PB for throughput improvement, we also proposes a hybrid mode scheme in this chapter. Different from the hybrid mode scheme defined in Chap. 3 where only the AB mode is adopted for information backscattering, both AB and BS modes are adopted in this chapter. That is to say, the CU can work in the HHT mode, the AB mode and the BS mode. When the primary channel is busy, the PT broadcasts RF signals, and the CU can harvest energy following the HTT mode or backscatter information in the AB mode. When the primary channel is idle, the PB broadcasts energy signals, the CU can still harvest energy or transmit information following the HTT mode, or backscatters information in the BS mode. To maximize the system throughput, we first formulate the optimization problem. Then, the closed-form solution is derived based on the KKT conditions and the optimal combination of the CU's working modes is revealed. Finally, numerical results are given to evaluate the performance superiority of the proposed hybrid mode scheme. Note that the majority of the contents of this chapter are based on our previous work [1].

4.2 System Model

As illustrated in Fig. 4.1, we study a CWPCN consisting of a primary communication pair and a secondary communication system. The primary communication pair includes a PT (e.g. TV tower) and a primary receiver (PR). The secondary communication system consists of a PB, an IR, and a CU, denoted by U_1. We assume that all terminals each has a single antenna. The CU is typically placed nearer to the PB than the PT. The PT and the PB have stable energy supplies, while the CU which does not have embedded energy sources need to be powered by the PT or the PB. The CU works in either the HTT mode or the BackCom mode, but not simultaneously [11]. When the PT transmits signals to the PR, i.e., the primary channel stays in the busy state, the CU can either harvest energy and store the harvested energy in their storages, or backscatter modulated signals to the IR. When the primary channel stays in the idle state, the PB is enabled to broadcast energy signals to the CU, during which the CU can also harvest energy or backscatter modulated signals. The harvested energy from both the PT and the PB is used for transmitting information signals to the IR.

The network is studied based on a time block with duration of T, the structure of which is illustrated in Fig. 4.2. Without loss of generality, we let $T = 1$. Denote the durations of the primary channel busy and idle periods as $1 - \beta$ and β, respectively. We assume the CU backscatters or transmits information via time division multiple access (TDMA). Denote the time allocated to U_1's BackCom mode during the primary channel busy and idle periods as α_1 and $\hat{\alpha}_1$, respectively. Moreover, during

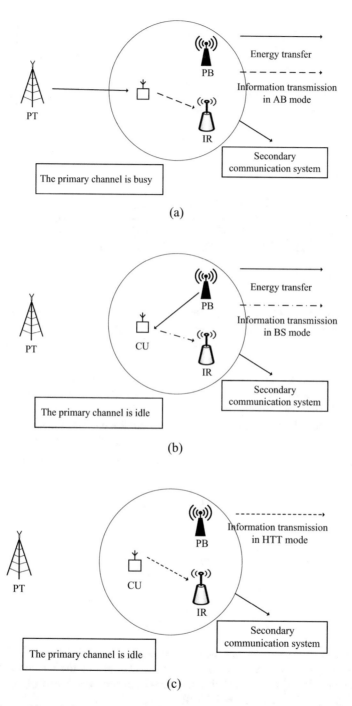

Fig. 4.1 A cognitive wireless powered communication network with hybrid BackCom. (**a**) CU harvests energy from the PT or backscatters information in the AB mode. (**b**) CU harvests energy from the PB or backscatters information in the BS mode. (**c**) CU transmits information in the HTT mode

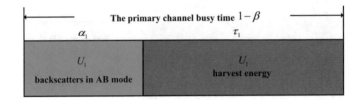

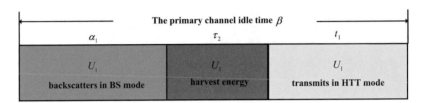

Fig. 4.2 Time block structure for a CWPCN

the primary channel busy and idle periods, the time slots with duration of τ_1 and τ_2 are respectively allocated to the CU for harvesting energy. During the primary channel idle period, the information transmission of the HTT mode is enabled. Denote the information transmission time of U_1 as t_1. It is assumed that the IR can apply the corresponding demodulators to extract useful information [3].

The HTT mode consists of two phases, i.e., WPT and WIT phases. That is to say, the CU first harvests energy from the PT or the PB, and then uses the harvested energy to transmit information to the IR. Denote the channel power gains between the PT and U_1, between the PB and U_1, and between U_1 and the IR as h_p^1, h_h^1 and g_h^1, respectively. We follow the Friis equation, h_p^1, h_h^1 and g_h^1 are thus given by

$$h_p^1 = \frac{G_p G_u \lambda_p^2}{(4\pi d_p^1)^2},$$

$$h_h^1 = \frac{G_h G_u \lambda_h^2}{(4\pi d_h^1)^2},$$

$$g_h^1 = \frac{G_u G_r \lambda_u^2}{(4\pi d_u^1)^2},$$

where G_p, G_h, G_u and G_r are the antenna gains of the PT, the PB, U_1, and the IR, respectively, λ_p, λ_h and λ_u are the signal wavelengths of the PT, the PB and U_1, respectively, d_p^1, d_h^1 and d_u^1 are the distances between the PT and U_1, between the PB and U_1, and between U_1 and the IR, respectively.

During the energy harvesting phase, the harvested energy of U_1 from the PT and the PB, denoted by E_p^1 and E_h^1, are given by

$$E_p^1 = \eta^1 P_p h_p^1 \tau_1,$$

$$E_h^1 = \eta^1 P_h h_h^1 \tau_2,$$

where P_p is the PT's transmit power, P_h is the PB's transmit power, η^1 is the energy harvesting efficiency of U_1.

During the information transmission phase, U_1 uses up its harvested energy to transmit information during t_1. Hence, the average transmit power of U_1 is given by

$$P_t^1 = \frac{E_p^1 + E_h^1}{t_1}.$$

The information transmission bits of U_1 in the HTT mode is given by

$$R_h^1 = t_1 W \log_2 (1 + \frac{P_t^1 g_h^1}{\sigma^2})$$

$$= t_1 W \log_2 \left(1 + \frac{\gamma_p^1 \tau_1 + \gamma_h^1 \tau_2}{t_1} \right), \qquad (4.1)$$

where W is the bandwidth, σ^2 is the noise power at the IR, $\gamma_p^1 = \frac{\eta^1 P_p h_p^1 g_h^1}{\sigma^2}$ and $\gamma_h^1 = \frac{\eta^1 P_h h_h^1 g_h^1}{\sigma^2}$.

We consider a hybrid BackCom that includes two types, e.g., AB and BS modes. Different from the HTT mode, for BackCom, the CU reflects the instantaneous incident signal from the PT or the PB to the IR such that the dedicated energy harvesting phase is not required. During the primary channel busy period, the PT serves as the incident signal source. Since the PT is far from the CUs, the AB mode can be activated. In contrary, the PB is placed nearer to the CUs than the PT, and the BS mode can thus be activated during the primary channel idle period.

Recently, many prototypes of AB transmission have been proposed [5, 6]. Here, we utilize the prototype designed in [5]. In this prototype, when the incident signal is available at a CU, the antenna impedance is varied to achieve the information backscattering, and the simple on-off keying (OOK) modulation is performed. It is reported that the backscatter rate is determined by the setting of the RC circuit elements [5]. Denote the backscatter rate of U_1 in the AB mode as B_a^1, and the backscatter bits of U_1 is thus given by

$$R_a^1 = B_a^1 \alpha_1. \qquad (4.2)$$

Since the backscattered signal power is very low, AB is unregulated by FCC [12]. Following this policy, we do not consider the interference caused by the AB mode at the primary communication pair. Moreover, the CUs can also offset the carrier phase by a certain frequency to avoid the interference [7].

When the BS mode is activated at U_1, the binary frequency-shift keying (FSK) modulation is performed following [8]. It is obvious that the IR can receive the radiated signal from the PB and the backscattered signal from CUs. Following [4], we assume that IR has the sophisticated preprocessing function such that it can compensate the carrier frequency offset and remove the direct current (DC) part of the received signal before the detection of the backscattered FSK signal [8]. Hence, the received power at the IR is expressed as

$$P_b^1 = P_h h_h^1 g_h^1 s^2 (\frac{\Gamma_0 - \Gamma_1}{2})^2 (\frac{4}{\pi})^2,$$

where s is the tag scattering efficiency, Γ_0 and Γ_1 are the reflection coefficients [4, 8]. The achievable rate of the BS mode is thus given by

$$B_b^1 = W \log_2(1 + \frac{\xi P_b^1}{\sigma^2}), \tag{4.3}$$

where ξ is the performance gap reflecting real modulation [4] (For more details, please refer to [4, 8]). From (4.3), we notice that the backscatter rate in the BS mode is controlled by both the circuit elements and the system parameters. Finally, the backscatter bits of U_1 in the BS mode is given by

$$R_b^1 = B_b^1 \hat{\alpha}_1. \tag{4.4}$$

Therefore, the total transmission bits of the secondary communication system is given by

$$R = R_h^1 + R_a^1 + R_b^1. \tag{4.5}$$

4.3 Throughput Maximization

Under the system model given in Sect. 4.2, we study the optimal time allocation strategy for maximizing the total transmission bits of the secondary communication system. Denote $\tau = [\tau_1, \tau_2]$. From (4.5), the optimization problem under the time allocation constraints is formulated as

$$\max_{\alpha_1, \hat{\alpha}_1, t_1, \boldsymbol{\tau}} \quad R(\alpha_1, \hat{\alpha}_1, t_1, \boldsymbol{\tau})$$

$$\text{s.t.} \quad \text{C1:} \quad \tau_1 + \alpha_1 \le 1 - \beta,$$
$$\text{C2:} \quad \tau_2 + \hat{\alpha}_1 + t_1 \le \beta, \qquad (4.6)$$
$$\text{C3:} \quad \tau_1, \tau_2 \ge 0,$$
$$\text{C4:} \quad \alpha_1, \hat{\alpha}_1, t_i \ge 0,$$

where C1 guarantees that the summation of the total backscattering time of U_1 in the AB mode and the dedicated energy harvesting time should not exceed the primary channel busy time, C2 limits the total time of the BS and HTT modes during the primary channel idle period, C3 and C4 are the non-negative constraints for the time allocation variables.

4.3.1 Multiple Cognitive Users Case

In this section, we study the optimal solution for (4.6). Similar as Lemma 2.1, we can prove that (4.6) is a convex optimization problem [9]. We obtain the combinations of working modes by analyzing the optimal solution.

Denote the optimal solution for (4.6) as α_1^*, $\hat{\alpha}_1^*$, t_1^* and $\boldsymbol{\tau}^* = [\tau_1^*, \tau_2^*]$. Before solving (4.6), we have the following lemma.

Lemma 4.1 *The optimal solution for (4.6) satisfies*

$$\tau_1^* + \alpha_1^* = 1 - \beta, \qquad (4.7)$$
$$\tau_2^* + \hat{\alpha}_1^* + t_1^* = \beta. \qquad (4.8)$$

Since $R(\alpha_1, \hat{\alpha}_1, \boldsymbol{\tau}, t_1)$ is an increasing function with respect to α_1 and $\hat{\alpha}_1$, the proof of Lemma 4.1 can be done by contradiction and is omitted here for simplicity.

To solve (4.6), we consider two cases following [4], i.e., $\tau_2 = 0$ and $0 < \tau_2 \le \beta$. If $\tau_2 = 0$, it indicates that U_1 can harvest sufficient energy when the primary channel stays in the busy state and the energy harvesting during the primary channel idle period is not necessary. While, if $\tau_2 > 0$, U_1 may harvest energy during both the primary channel busy and idle periods. From Lemma 4.1, Problem P2 for the case $\tau_2 = 0$ is rewritten as

$$\max_{\tau_1, t_1} \quad R(\tau_1, t_1)$$

$$\text{s.t.} \quad 0 \le \tau_1 \le 1 - \beta, \qquad (4.9)$$
$$0 \le t_1 \le \beta,$$

where $R(\tau_1, , t_1) = t_1 W \log_2 \left(1 + \frac{\gamma_p^1 \tau_1}{t_1}\right) + B_a^1(1 - \beta - \tau_1) + B_b^1(\beta - t_1)$.

It is obvious that (4.9) is a convex optimization problem. The Lagrangian is given by

$$\mathcal{L}(\tau_1, t_1) = t_1 W \log_2 \left(1 + \frac{\gamma_p^1 \tau_1}{t_1}\right) + B_a^1(1 - \beta - \tau_1)$$

$$+ B_b^1(\beta - t_1) + \mu_1 \tau_1 - \mu_2(\tau_1 - 1 + \beta)$$

$$+ \nu_1 t_1 - \nu_2(t_1 - \beta), \tag{4.10}$$

where μ_i and ν_i, $i = 1, 2$, are the Lagrangian multipliers. From (4.9), we observe that there exists a solution $\{\tau_1, t_1\}$ satisfying the constraints $0 < \tau_1 < 1 - \beta$ and $0 < t_1 < \beta$, which indicates that strong duality holds since the Slater's condition is satisfied [9]. Hence, (4.9) can be solved by KKT conditions, which are given by

$$\frac{\partial \mathcal{L}}{\partial t_1} = W \log_2(1 + \frac{\gamma_p^1 \tau_1^*}{t_1^*}) - W \frac{\frac{\gamma_p^1 \tau_1^*}{t_1^*}}{\ln 2(1 + \frac{\gamma_p^1 \tau_1^*}{t_1^*})}$$

$$- B_b^1 + \nu_1^* - \nu_2^* = 0, \tag{4.11}$$

$$\frac{\partial \mathcal{L}}{\partial \tau_1} = \frac{W \gamma_p^1}{\ln 2(1 + \frac{\gamma_p^1 \tau_1^*}{t_1^*})} - B_a^1 + \mu_1^* - \mu_2^* = 0, \tag{4.12}$$

$$\mu_1^* \tau_1^* = 0, \tag{4.13}$$

$$\mu_2^*(\tau_1^* - 1 + \beta) = 0, \tag{4.14}$$

$$\nu_1^* t_1^* = 0, \tag{4.15}$$

$$\nu_2^*(t_1^* - \beta) = 0, \tag{4.16}$$

where τ_1^*, t_1^*, μ_i^* and ν_i^*, $i = 1, 2$, are the optimal solution for (4.9) Before analyzing the details of the KKT conditions, we first give Lemma 4.2 as follows.

Lemma 4.2 *There exists a unique $z^* > 2^{\frac{B_b^1 + c}{W}}$ satisfies $f(z, c) = 0$, where*

$$f(z, c) = z \ln z - \left(1 + \frac{\ln 2(B_b^1 + c)}{W}\right) z + 1,$$

and c is a constant.

Proof The first and second derivatives of $f(z, c)$ are given by

$$\frac{\partial f(z, c)}{\partial z} = \ln z - \ln 2^{\frac{B_b^1 + c}{W}}, \tag{4.17}$$

$$\frac{\partial^2 f(z, c)}{\partial z^2} = \frac{1}{z}. \tag{4.18}$$

From (4.17) and (4.18), we derive the minimum of $f(z, c)$ is achieved at $z = 2^{\frac{B_b^1 + c}{W}}$ with $f(z = 2^{\frac{B_b^1 + c}{W}}, c) = 1 - 2^{\frac{B_b^1 + c}{W}} < 0$, and $f(z, c)$ is an increasing function with $z > 2^{\frac{B_b^1 + c}{W}}$. Moreover, we have

$$\lim_{z \to +\infty} f(z, c) > 0. \tag{4.19}$$

Hence, we derive that there is a unique solution for $f(z, c) = 0$ with $z > 2^{\frac{B_b^1 + c}{W}}$.

Lemma 4.3 *If* $\frac{\ln 2(B_a^1 - B_b^1)}{W} < \gamma_p^1$, *there exists a unique* $z^\dagger > 1$ *satisfies* $g(z) = \gamma_p^1$, *where*

$$g(z) = f(z, c = 0) + \frac{\ln 2 B_a^1}{W} z.$$

The proof of Lemma 4.3 is similar as Proof 4.3.1 and is omitted for simplicity.

For the KKT conditions, we first consider the following five cases with $\mu_1^* = 0$, which corresponds to $\tau_1^* \geq 0$.

Case 1: $\mu_2^* > 0$, $v_1^* = 0$ and $v_2^* = 0$. From (4.14), we derive that $\tau_1^* = 1 - \beta$. Based on (4.11), we have the following equation

$$f(z^*, c) = 0,$$

where $z^* = 1 + \frac{\gamma_p^1 \tau_1^*}{t_1^*}$ is its unique solution with $c = 0$ following Lemma 4.2. t_1^* is thus expressed as $t_1^* = \frac{\gamma_p^1 \tau_1^*}{z^* - 1} = \frac{\gamma_p^1 (1 - \beta)}{z^* - 1}$. This optimal solution indicates that U_1 will not work in the AB mode, but it first enters the BS mode with $\beta - \frac{\gamma_p^1 (1 - \beta)}{z^* - 1}$ amount of time and then switches to the information transmission phase of the HTT mode. We then analyze the conditions for case 1. Based on $t_1^* \leq \beta$, we have the condition $z^* \geq x^*$, where $x^* = 1 + \frac{\gamma_p^1 (1 - \beta)}{\beta}$. Moreover, from (4.12), we have the condition $B_a^1 < \frac{W \gamma_p^1}{\ln 2 z^*}$. The two conditions for case 1 can be combined as $B_a^1 < \frac{W \gamma_p^1}{\ln 2 z^*} \leq \frac{W \gamma_p^1}{\ln 2 x^*}$. Note that a similar analysis can be found in [4].

Case 2: $\mu_2^* > 0$, $v_1^* = 0$ and $v_2^* > 0$. From (4.14) and (4.16), we derive that $\tau_1^* = 1 - \beta$ and $t_1^* = \beta$. From the optimal solution, we observe that both AB and BS modes cannot be activated and the secondary communication system works as a WPCN proposed in [10]. The conditions of case 2 are given as follows. We have the following conditions from (4.11) and (4.12): $z^+ > z^*$ and $B_a^1 < \frac{W\gamma_p^1}{\ln 2z^+}$, where z^+ is the unique solution of $f(z, c) = 0$ with $c = v_2^* > 0$. We also derive that $z^+ = 1 + \frac{\gamma_p^1(1-\beta)}{\beta} = x^*$. The conditions are combined as $B_a^1 < \frac{W\gamma_p^1}{\ln 2x^*} < \frac{W\gamma_p^1}{\ln 2z^*}$.

Case 3: $\mu_2^* = 0$, $v_1^* = 0$ and $v_2^* > 0$. From (4.16), we derive $t_1^* = \beta$. From (4.12), we have $\tilde{z} = \frac{W\gamma_p^1}{\ln 2B_a^1}$, where $\tilde{z} = 1 + \frac{\gamma_p^1 \tau_1^*}{t_1^*}$. Hence, we derive $\tau_1^* = \frac{(\tilde{z}-1)\beta}{\gamma_p^1} = \frac{W\beta}{\ln 2B_a^1} - \frac{\beta}{\gamma_p^1}$. From the result, we have the following observation. When the primary channel is busy, U_1 first works in the AB mode with $1 - \beta - \frac{W\beta}{\ln 2B_a^1} + \frac{\beta}{\gamma_p^1}$ amount of time and then harvests energy with the remaining time. In contrary, the BS mode cannot be activated when the primary channel stays in the idle state. Since $\tau_1^* \leq 1 - \beta$, we have the condition $\tilde{z} \leq x^*$. Furthermore, we also have the condition $\tilde{z} > z^*$ as given in case 2 according to (4.11). The two conditions are combined as $\frac{W\gamma_p^1}{\ln 2x^*} \leq B_a^1 < \frac{W\gamma_p^1}{\ln 2z^*}$.

Case 4: $\mu_2^* = 0$, $v_1^* = 0$ and $v_2^* = 0$. According to Lemma 4.3, if we can find a solution $z^\dagger$ satisfying

$$f(z, c) = \gamma_p^1 - \frac{\ln 2B_a^1}{W} z \tag{4.20}$$

with $c = 0$, we have $z^\dagger = z^* = \tilde{z}$. $\{\tau_1^*, t_1^*\}$ can be any values in the feasible domain satisfying $z^\dagger = 1 + \frac{\gamma_p^1 \tau_1^*}{t_1^*}$. The condition of case 4 can be given as $B_a^1 = \frac{W\gamma_p^1}{\ln 2z^*}$. If there does not exist a solution for (4.20), this case does not exist.

Case 5: $\mu_2^* = 0$, $v_1^* > 0$ and $v_2^* = 0$. From (4.15), we have $t_1^* = 0$, which indicates that the information transmission phase of the HHT mode cannot work. Hence, the energy harvesting is not necessary, i.e., $\tau_1^* = 0$. From the optimal solution, we know that whether the primary channel is busy or idle, U_1 always stays in the BackCom mode. The condition for this case is given as follows. From (4.12), we have $B_a^1 = \frac{W\gamma_p^1}{\ln 2\check{z}}$ and $\check{z} < z^*$, where $\check{z}$ is the optimal solution of $f(z, c) = 0$ with $c = -v_1^* < 0$. Hence, the condition for this case is given as $B_a^1 > \frac{W\gamma_p^1}{\ln 2z^*}$.

Next, we consider the following case with $\mu_1^* > 0$, which corresponds to $\tau_1^* = 0$ according to (4.13).

Case 6: $\mu_2^* = 0$, $v_1^* \geq 0$ and $v_2^* = 0$. Since $\tau_1^* = 0$, U_1 cannot harvest any energy from the PT. Hence, U_1 cannot transmit any data, i.e., $t_1^* = 0$. This case is similar as case 5, i.e., the HTT mode cannot be activated and U_1 works in the BackCom mode over the whole block. The condition for this case is given as follows. We also denote $\check{z}$ as the optimal solution of $f(z, c) = 0$ with $c = -v_1^* \leq$

0. From (4.12), we have $B_a^1 > \frac{W\gamma_p^1}{\ln 2\breve{z}}$ and $\breve{z} \leq z^*$. The condition is thus given as $B_a^1 > \frac{W\gamma_p^1}{\ln 2z^*}$. Hence, case 5 and case 6 can be combined.

Note that under $\mu_1^* \geq 0$, there are still other combinations of the Lagrangian multipliers. However, the other combinations make the KKT conditions contradict with each other. Hence, we do not discuss the cases corresponding to the unmentioned combinations here.

We then consider (4.6) with $\tau_2 > 0$. From Lemma 4.1, (4.6) is recast as (4.21)

$$\max_{\boldsymbol{\tau},t_1} \quad R(\boldsymbol{\tau}, t_1)$$
$$\text{s.t.} \quad \tau_1 \leq 1 - \beta,$$
$$\tau_2 + t_1 \leq \beta, \tag{4.21}$$
$$\tau_2, t_1 > 0, \ \tau_1 \geq 0.$$

where $R(\boldsymbol{\tau}, t_1) = t_1 W \log_2\left(1 + \frac{\gamma_p^1 \tau_1 + \gamma_h^1 \tau_2}{t_1}\right) + B_a^1(1 - \beta - \tau_1) + B_b^1(\beta - t_1 - \tau_2)$.

Equation (4.21) is also a convex optimization problem. Before solving (4.21), we first derive the optimal relationship between τ_2 and t_1 with given τ_1. Let $\bar{t} = \tau_2 + t_1 \leq \beta$. Hence, (4.21) is recast as (4.22), which is given by

$$\max_{\tau_2,t_1} \quad R(\tau_2, t_1)$$
$$\text{s.t.} \quad t_1 + \tau_2 = \bar{t}, \tag{4.22}$$
$$t_1, \tau_2 > 0.$$

Equation (4.22) can also be solved by KKT conditions and its Lagrangian is given as

$$\mathcal{L}(\tau_2, t_1) = t_1 W \log_2\left(1 + \frac{\gamma_p^1 \tau_1 + \gamma_h^1 \tau_2}{t_1}\right)$$
$$+ B_a^1(1 - \beta - \tau_1) + B_b^1(\beta - t_1 - \tau_2)$$
$$- \zeta(t_1 + \tau_2 - \bar{t}),$$

where ζ is the Lagrangian multiplier. The KKT conditions are given by

$$\frac{\partial\mathcal{L}}{\partial\tau_2} = W\frac{\gamma_h^1}{\ln 2(1 + \frac{\gamma_p^1\tau_1^* + \gamma_h^1\tau_2^*}{t_1^*})} - B_b^1 - \zeta^* = 0, \tag{4.23}$$

$$\frac{\partial \mathscr{L}}{\partial t_1} = W \log_2(1 + \frac{\gamma_p^1 \tau_1^* + \gamma_h^1 \tau_2^*}{t_1^*}) - W \frac{\frac{\gamma_p^1 \tau_1^* + \gamma_h^1 \tau_2^*}{t_1^*}}{\ln 2(1 + \frac{\gamma_p^1 \tau_1^* + \gamma_h^1 \tau_2^*}{t_1^*})}$$

$$- B_b^1 - \zeta^* = 0, \tag{4.24}$$

$$\zeta^*(t_1^* + \tau_2^* - \bar{t}) = 0, \tag{4.25}$$

Lemma 4.4 is given here for the following discussions of the KKT conditions.

Lemma 4.4 *There exists a unique* $\bar{z} > 1$ *satisfies* $h(z) = b$ $(b > 0)$, *where*

$$h(z) = z \ln z - z + 1.$$

The proof of Lemma 4.4 is also omitted for simplicity.
From (4.23) and (4.24), we derive τ_2^* and t_1^*, which are given by

$$\tau_2^* = \frac{\bar{z}\bar{t} - \bar{t} - \gamma_p^1 \tau_1}{\bar{z} - 1 + \gamma_h^1}, \tag{4.26}$$

$$t_1^* = \frac{\bar{t}\gamma_h^1 + \gamma_p^1 \tau_1}{\bar{z} - 1 + \gamma_h^1}, \tag{4.27}$$

where $\bar{z} > 1$ is the unique solution of $h(z) = \gamma_h^1$. With (4.26) and (4.27), we have
$R(\bar{t}, \tau_1) = \frac{\bar{t}\gamma_h^1 + \gamma_p^1 \tau_1}{\bar{z}-1+\gamma_h^1} W \log_2 \bar{z} + B_a^1(1 - \beta - \tau_1) + B_b^1(\beta - \bar{t}) = (\frac{\gamma_h^1}{\bar{z}-1+\gamma_h^1} W \log_2 \bar{z} -$
$B_b^1)\bar{t} + (\frac{\gamma_p^1}{\bar{z}-1+\gamma_h^1} W \log_2 \bar{z} - B_a^1)\tau_1 + B_a^1(1-\beta) + B_b^1\beta$. Equation (4.21) is reformulated
as

$$\max_{\tau_1, \bar{t}} \quad R(\bar{t}, \tau_1)$$
$$\text{s.t.} \quad 0 < \bar{t} \leq \beta, \tag{4.28}$$
$$0 \leq \tau_1 \leq 1 - \beta.$$

Denote the optimal solution for (4.28) as τ_1^* and $\bar{t}^*$. To solve (4.28), we consider the
following two cases by taking $\bar{t} > 0$ into account.

Case 7: $\frac{\gamma_h^1}{\bar{z}-1+\gamma_h^1} W \log_2 \bar{z} > B_b^1$ and $\frac{\gamma_p^1}{\bar{z}-1+\gamma_h^1} W \log_2 \bar{z} \geq B_a^1$. In this case, we derive
that $\bar{t}^* = \beta$ and $\tau_1^* = 1-\beta$. Hence, the other optimal variables for Problem P4 are
given as $\tau_2^* = \frac{\bar{z}\beta - \beta - \gamma_p^1(1-\beta)}{\bar{z}-1+\gamma_h^1}$ and $t_1^* = \frac{\beta\gamma_h^1 + \gamma_p^1(1-\beta)}{\bar{z}-1+\gamma_h^1}$. For this case, the secondary
communication system only works in the HTT mode. Since $0 < \tau_2 < \beta$, we
further have the condition $\bar{z} > 1 + \frac{\gamma_p^1(1-\beta)}{\beta}$.

Table 4.1 Optimal solution of (4.6)

Condition		Optimal solution
$\tau_2^* = 0$	$B_a^1 < \frac{W\gamma_p^1}{\ln 2z^*} \leq \frac{W\gamma_p^1}{\ln 2x^*}$	$\tau_1^* = 1 - \beta,\ t_1^* = \frac{\gamma_p^1(1-\beta)}{z^*-1}$
	$B_a^1 < \frac{W\gamma_p^1}{\ln 2x^*} < \frac{W\gamma_p^1}{\ln 2z^*}$	$\tau_1^* = 1 - \beta,\ t_1^* = \beta$
	$\frac{W\gamma_p^1}{\ln 2x^*} \leq B_a^1 < \frac{W\gamma_p^1}{\ln 2z^*}$	$\tau_1^* = \frac{W\beta}{\ln 2 B_a^1} - \frac{\beta}{\gamma_p^1},\ t_1^* = \beta$
	$B_a^1 > \frac{W\gamma_p^1}{\ln 2z^*}$	$\tau_1^* = 0,\ t_1^* = 0$
	$B_a^1 = \frac{W\gamma_p^1}{\ln 2z^*}$	Any feasible values
$\tau_2^* > 0$	$\frac{\gamma_h^1}{\bar{z}-1+\gamma_h^1} W \log_2 \bar{z} > B_b^1,$	$\tau_1^* = 1 - \beta,$
	$\frac{\gamma_p^1}{\bar{z}-1+\gamma_h^1} W \log_2 \bar{z} \geq B_a^1,$	$t_1^* = \frac{\beta\gamma_h^1 + \gamma_p^1(1-\beta)}{\bar{z}-1+\gamma_h^1},$
	$\bar{z} > 1 + \frac{\gamma_p^1(1-\beta)}{\beta}$	$\tau_2^* = \frac{\bar{z}\beta - \beta - \gamma_p^1(1-\beta)}{\bar{z}-1+\gamma_h^1}$
	$\frac{\gamma_h^1}{\bar{z}-1+\gamma_h^1} W \log_2 \bar{z} > B_b^1,$	$\tau_1^* = 0,\ t_1^* = \frac{\beta\gamma_h^1}{\bar{z}-1+\gamma_h^1},$
	$\frac{\gamma_p^1}{\bar{z}-1+\gamma_h^1} W \log_2 \bar{z} < B_a^1$	$\tau_2^* = \frac{\bar{z}\beta - \beta}{\bar{z}-1+\gamma_h^1}$

Case 8: $\frac{\gamma_h^1}{\bar{z}-1+\gamma_h^1} W \log_2 \bar{z} > B_b^1$ and $\frac{\gamma_p^1}{\bar{z}-1+\gamma_h^1} W \log_2 \bar{z} < B_a^1$. In this case, we derive that $\bar{t}^* = \beta$ and $\tau_1^* = 0$. Hence, the other optimal variables for Problem P4 are given as $\tau_2^* = \frac{\bar{z}\beta - \beta}{\bar{z}-1+\gamma_h^1}$ and $t_1^* = \frac{\beta\gamma_h^1}{\bar{z}-1+\gamma_h^1}$. For this case, U_1 works in the AB mode when the primary channel stays in the busy state and works in the HTT mode when the primary channel stays in the idle state.

The optimal solutions of (4.6) is summarized in Table 4.1. The optimal combination of working modes corresponds to the result that achieves the maximum throughput.

4.4 Numerical Results and Discussions

In this section, numerical simulations are conducted to evaluate the performance of the proposed hybrid HTT-BackCom mode. We also present the numerical results of the benchmark for comparison. In our numerical experiments, the PT transmits signals at 915 MHz with 6 MHz bandwidth, while the PB transmits signals at 2.4 GHz with 20 MHz bandwidth. The antenna gain of each CU is set as 1.8 dBi, the antenna gains of other devices are set as 6 dBi, and the reflection coefficients of the BS mode are set as $\Gamma_0 = 1$ and $\Gamma_1 = -1$, respectively. The scattering efficiency s causes a power loss of 1.1 dB. Unless otherwise stated, we set $P_p = 17$ kW, $P_h = 20$ dBm, $d_p^1 = 1$ km, $d_h^1 = 1.5$ m, $d_u^1 = 2$ m, $\beta = 0.3$, $\eta^1 = 0.6$, $\xi = -5$ dB, $\sigma^2 = -20$ dBm, $B_a^1 = 3$ Mbps. We show that our proposed hybrid HTT-BackCom mode outperforms the traditional HTT mode in terms of system throughput.

Figure 4.3 shows the throughput versus the primary channel idle time. From Fig. 4.3, we can see that the proposed hybrid HTT-BackCom mode leads to a larger

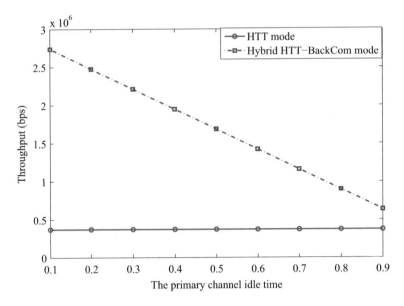

Fig. 4.3 Throughput vs. primary channel idle time

throughput than the HTT mode. This can be explained as follows. Under the given parameters, U_1 always works in the AB mode when the primary channel stays in the busy state and works in the HTT mode when the primary channel stays in the idle state for the proposed hybrid HTT-BackCom mode. Moreover, the AB mode's backscatter rate is larger than the HTT mode's transmission rate. We also see that the throughput achieved by the proposed hybrid HTT-BackCom mode decreases with the increase of the primary channel idle time β. It is because as β increases, the time for the AB mode reduces and the time for the HTT mode increases, which reduces the throughput. In Fig. 4.4, we vary the AB mode's backscatter rate to evaluate the throughput. It is obvious that the AB mode affects the throughput significantly.

Figure 4.5 shows the throughput of the secondary communication system versus the PB's transmit power. We observe that the throughput is an increasing function with respect to the PB's transmit power for both the proposed hybrid HTT-BackCom mode and HTT mode. As the transmit power increases, the throughput increases first slowly and then fast. This can be explained as follows. The rates of both BS and HTT modes increase with the increase of the PB's transmit power. When the PB's transmit power is low, U_1 works in the HTT mode for information transmission during the primary channel idle period. When the PB's transmit power exceeds a threshold (e.g., 30 dBm), U_1 works in the BS mode for information backscattering and the rate of the BS mode increases fast.

Figures 4.6 and 4.7 show the throughput versus the distances between the PB and U_1 and between the IR and U_1, respectively. It is obvious that as the distances increase, the throughput reduces fast first and then slowly. The observation in

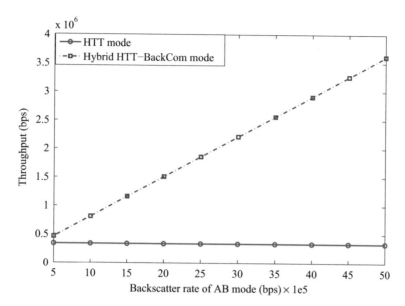

Fig. 4.4 Throughput vs. AB mode's rate

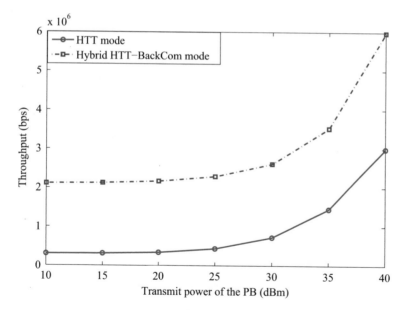

Fig. 4.5 Throughput vs. PB's transmit power

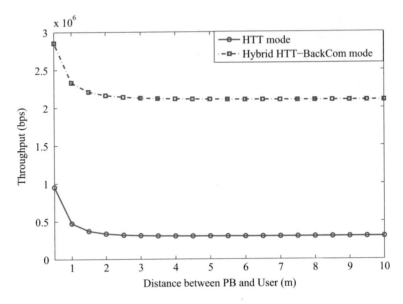

Fig. 4.6 Throughput vs. distance between PB and user

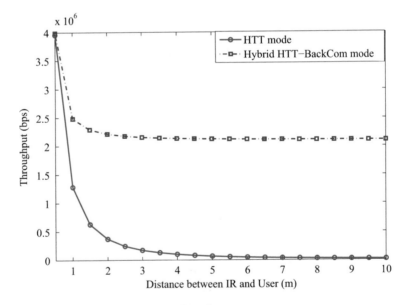

Fig. 4.7 Throughput vs. distance between IR and user

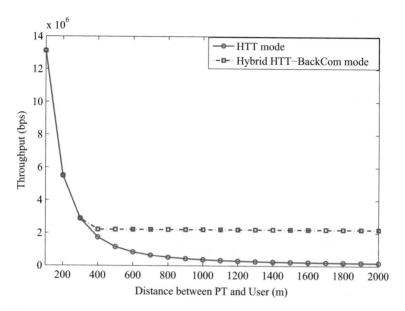

Fig. 4.8 Throughput vs. distance between PT and user

Fig. 4.6 is because the energy harvested from the PB first reduces significantly and then tends to be negligible. In Fig. 4.7, when the distance $d_u^1 = 0.5$ m, the throughput of the proposed and benchmark modes is almost the same. It is because in this case the HTT mode dominates the throughput for the proposed mode.

Figure 4.8 depicts the curves of the throughput versus the distance between the PT and U_1. Similar with Figs. 4.6 and 4.7, the throughput reduces as the distance increases. When the distance is small, the throughput achieved by the proposed hybrid HTT-BackCom mode and benchmark mode is the same since U_1 only works in the HTT mode for the proposed mode. When the distance exceeds a threshold (e.g., 300 m), U_1 first works in the HTT mode and then in the AB mode due to the reduce of harvested energy.

4.5   Conclusions

In this paper, we have proposed a novel hybrid HTT-BackCom mode in CWPCNs, where the CU in the secondary communication system work in the HTT mode, the AB mode or the BS mode. The signal from the PT and the signal from the PB serve as the excitation signals for the AB mode and BS mode, respectively, while the PT and PB are both the energy sources for the HTT mode. Under the proposed hybrid mode, we have investigated the secondary communication system throughput maximization problem by finding the optimal time allocation between the AB mode and energy harvesting when the primary channel stays in the busy state and that between the BS mode and the HTT mode when the primary channel stays in the idle state. We have derived the optimal solution in closed-form and obtained the optimal combination of the working modes. Numerical results have confirmed that our proposed hybrid HTT-BackCom achieves a higher throughput than the traditional HTT mode, which shows the advantages of combining the HTT and BackCom modes.

References

1. Bin Lyu, Haiyan Guo, Zhen Yang, and Guan Gui, "Throughput maximization for hybrid backscatter assisted cognitive wireless powered communication networks," *IEEE Internet of Things Journal*, https://doi.org/10.1109/JIOT.2018.2820180.
2. S. Lee, R. Zhang, "Cognitive wireless powered network: Spectrum sharing models and throughput maximization," *IEEE Trans. Cogn. Commun. Netw.*, vol. 1, no. 3, pp. 335–346, Sep. 2015.
3. D. T. Hoang, D. Niyato, P. Wang, D. I. Kim, and Z. Han, "The tradeoff analysis in RF-powered backscatter cognitive radio networks," in *IEEE GLOBECOM*, Washington, USA, Dec. 2016, pp. 1–6.
4. S. H. Kim and D. I. Kim, "Hybrid backscatter communication for wireless-powered heterogeneous networks," *IEEE Trans. Wireless Commun.*, vol. 16, no. 10, pp. 6557–6570, Jul. 2017.
5. V. Liu, A. Parks, V. Talla, S. Gollakota, D. Wetherall, and J. R. Smith, "Ambient backscatter: Wireless communication out of thin air," in *Proc. SIGCOMM*, pp. 39–50, Hong Kong, Aug. 2013.
6. A. N. Parks, A. Liu, S. Gollakota, and J. R. Smith, "Turbocharging ambient backscatter communication," in *Proc. SIGCOMM*, pp. 619–630, Chicago, USA, Aug. 2014.
7. B. Kellogg, V. Talla, S. Gollakota, and J. R. Smith, "Passive wi-fi: Bringing low power to wi-fi transmissions," in *Proc. NSDI*, pp. 151–164, Santa Clara, USA, Mar. 2016.
8. J. Kimionis, A. Bletsas, and J. N. Sahalos, "Increased range bistatic scatter radio," *IEEE Trans. Commun.*, vol. 62, no. 3, pp. 1091–1104, Mar. 2014.
9. S. Boyd and L. Vandenberghe, *Convex Optimization*. Cambridge University Press, 2004.
10. H. Ju and R. Zhang, "Throughput maximization in wireless powered communication networks," *IEEE Trans. Wireless Commun.*, vol. 13, no. 1, pp. 418–428, Jan. 2014.
11. P. Zhang and D. Ganesan, "Enabling bit-by-bit backscatter communication in severe energy harvesting environment," *in Proc. NSDI*, pp. 345–357, Seattle, WA, Apr. 2014.
12. New policies for part 15 devices, FCC, TCBC workshop, 2005.

Chapter 5
Relay Cooperation for Backscatter Communication Systems

Abstract A relay cooperation scheme is proposed for a backscatter communication system (BCS), where the user backscatters incident signals from a carrier emitter (CE) to an IR and a relay simultaneously, and then the relay forwards the user's information to the IR for throughput improvement. Two cases are considered that the relay is with an embedded energy source and the relay is without an embedded energy source. If the relay does not have an embedded energy source, it first harvests energy from the CE and then uses its harvested energy for information forwarding. For both cases, the time allocation problems on the user's information backscattering, the user's information forwarding, or the relay's energy harvesting are formulated to maximize the system throughput, and then closed-form solutions are derived. Simulation results demonstrate the advantage of the proposed relay cooperation scheme with the optimal time allocation in terms of system throughput.

5.1 Introduction

The HTT mode is widely used in WPCNs [2]. The users following the HTT mode transmit information actively, which usually require the oscillators to generate carrier signals and analog-to-digital converters (ADCs) for digital modulation. Hence, the circuit power consumption of these devices is not low enough for the long-lifetime requirement for IoT. The backscatter communication system (BCS) can be seen as a special case of WPCNs, where the BackCom mode is adopted. In a BCS, the users do not transmit information actively but reflect and modulate the incident signals via mismatching the devices' antenna impedance [4]. Hence, the active RF components are not necessary and the circuit power consumption of the users in the BackCom mode is orders-of-magnitude less than the users in the HTT mode.

In the BCS, due to the path-loss between the CE and the users and between the users and the IR, the power of the received signal at the IR is typically small, which limits the system throughput. The cooperative transmission technique has been extensively applied in wireless communication systems to increase system capacity.

© The Author(s), under exclusive license to Springer Nature Switzerland AG 2019
G. Gui, B. Lyu, *Optimization for Wireless Powered Communication Networks*,
SpringerBriefs in Electrical and Computer Engineering,
https://doi.org/10.1007/978-3-030-01021-8_5

In a cooperative network, the relay node is usually used to forward information transmission [5]. Inspired by this, in this chapter, we employ a relay to improve the throughput of BCSs. In the considered system, the user backscatters the continuous carrier wave (CW) from the CE to both the relay and the receiver simultaneously, and then the relay forwards the received signal from the user to the receiver. We consider two cases that the relay is with an embedded energy source and the relay is not with an embedded energy source. For the first case, the relay uses its own energy for information forwarding. While for the latter case, the relay needs to harvest energy from the CE first and then uses its harvested energy to forward the user's information. For both cases, we study the system throughput maximization and derive the optimal time allocation schemes respectively. Numerical results show that the proposed relay cooperation scheme can significantly improve the system throughput. Note that the majority of the contents of this chapter are based on our previous work [1].

5.2 System Model

As illustrated in Fig. 5.1, we study a BCS, where each terminal has one single antenna. The system is studied based on a transmission block with duration of T. Denote the distances between the CE and the user, between the CE and the relay, between the CE and the receiver, between the user and the receiver, between the user and the relay, and between the relay and the receiver as $d_{0,1}$, $d_{0,2}$, $d_{0,3}$, d_1, d_2 and d_3. To make the relay assist the information transmission between the user and the receiver, i.e., the relay first decodes the information backscattered by the user and then forwards the decoded information to the receiver, we assume that $d_1 > d_2$ and $d_1 > d_3$. Denote the channel gains between the CE and the user, between the CE and the relay, between the CE and the receiver, between the user and the receiver, between the user and the relay, and between the relay and the receiver as $h_{0,1}$, $h_{0,2}$, $h_{0,3}$, h_1, h_2, and h_3, which are modelled as quasi-static flat-fading and remain constant during each transmission block, but may vary from one block to another. Following [3], we only consider the distance-dependent signal attenuation such that we have $h_1 < h_2$ and $h_1 < h_3$.

We assume the CE transmits the continuous CW with power P and denote its transmitted signal as s, where $\mathbb{E}[|s|^2] = P$. The user backscatters information based on the incident signal from the CE. We consider two cases for the relay, i.e., the relay is with an embedded energy source and the relay is without an embedded energy source. To simplify the aftermentioned descriptions, we denote the cases that the relay is with/without an embedded energy source as Case A and Case B, respectively.

Fig. 5.1 A RaBackCom
system

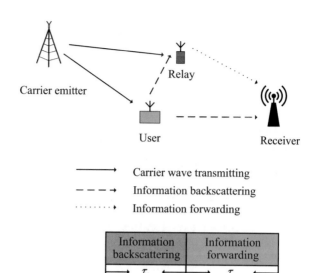

Carrier emitter

Relay

User

Receiver

——————→   Carrier wave transmitting

– – – →   Information backscattering

........→   Information forwarding

Fig. 5.2 Block structure. (**a**)
Case A. (**b**) Case B

Information backscattering	Information forwarding
τ_1	τ_2

(a)

Energy Harvesting	Information backscattering	Information forwarding
t_0	t_1	t_2

(b)

5.2.1   Relay with an Embedded Energy Source

If the relay has an embedded energy source, it does not need to harvest energy
from the CE. The structure of a transmission block is illustrated in Fig. 5.2a. During
τ_1, the CE transmits the CW signal and the received signal at the user is given by
$\sqrt{h_{0,1}}s$ without considering the noise following [6]. Denote the reflection coefficient
at the user as α, where $0 \leq \alpha \leq 1$. In this paper, we assume that the main energy
consumption of the user and the relay is for information transmission and does not
consider other circuit energy consumption for simplicity. Hence, we assume that
$\alpha = 1$, i.e., all the received signal at the user will be backscattered. Following the
definition in [7], the user's own signal is denoted as c, where $\mathbb{E}[|c|^2] = 1$. The
backscattered signal at the user, denoted by x, is thus expressed as

$$x = \alpha\sqrt{h_{0,1}}sc = \sqrt{h_{0,1}}sc, \qquad (5.1)$$

and the received signals at the receiver and the relay, denoted as y_1 and y_2, are
respectively given by

$$y_1 = \sqrt{h_1}\sqrt{h_{0,1}}sc + \sqrt{h_{0,3}}s + n_1, \qquad (5.2)$$

$$y_2 = \sqrt{h_2}\sqrt{h_{0,1}}sc + \sqrt{h_{0,2}}s + n_2. \tag{5.3}$$

where $n_i, i = 1, 2$, is the Gaussian noise satisfying $n_i \sim \mathcal{CN}(0, \sigma^2)$, $\sqrt{h_{0,3}}s$ and $\sqrt{h_{0,2}}s$ are the interference signals from the CE at the receiver and the relay, respectively. We assume that successive interference cancellation (SIC) is adopted at both the receiver and the relay following [7]. Hence, both the receiver and the relay can first decode the interference signals and then subtract them from the received signals. Denote the signal-noise-ratio (SNR) at the receiver and the relay during τ_1 as γ_1 and γ_2, respectively, which are given by $\gamma_1 = \frac{Ph_{0,1}h_1}{\sigma^2}$ and $\gamma_2 = \frac{Ph_{0,1}h_2}{\sigma^2}$.

Denote the relay's energy for forwarding the user's information as E, which is exhausted during τ_2. Hence, the average transmit power at the relay is given by $P_r = \frac{E}{\tau_2}$. Denote the forwarded signal at the relay as $\hat{x}$, where $\mathbb{E}[|\hat{x}|^2] = P_r$. The received signal at the receiver from the relay is thus expressed as

$$y_3 = \sqrt{h_3}\hat{x} + n_1. \tag{5.4}$$

Denote the instantaneous transmission rates from the user to the receiver, from the user to the relay, from the relay to the receiver as R_1, R_2 and R_3, respectively, which are expressed as

$$R_1 = \tau_1 \log_2(1 + \gamma_1), \tag{5.5}$$

$$R_2 = \tau_1 \log_2(1 + \gamma_2), \tag{5.6}$$

$$R_3 = \tau_2 \log_2(1 + \frac{\gamma_3}{\tau_2}), \tag{5.7}$$

where $\gamma_3 = \frac{h_3 E}{\sigma^2}$. From [8], the instantaneous transmission rate of the user, denoted by R, is given by

$$R = \min\{R_1 + R_3, R_2\}. \tag{5.8}$$

5.2.2   Relay Without an Embedded Energy Source

If the relay does not have an embedded energy source, it needs to harvest energy from the CE. The block structure of this case is given in Fig. 5.2b. Different from Case A, an energy harvesting time for the relay is required here. During t_0, we assume that only the relay is activated to harvest energy from the CE for simplicity, where the harvested energy is given by

$$\hat{E} = \eta Ph_{0,2}t_0,$$

where η is the energy harvesting efficiency. During t_1 and t_2, the system works as that we describe during τ_1 and τ_2 in Sect. 5.2.1. Hence, the average transmit power for the relay during t_2, denoted by $\hat{P}_r$, is given by $\hat{P}_r = \frac{\eta P h_{0,2} t_0}{t_2}$.

Denote the instantaneous transmission rates from the user to the receiver, from the user to the relay, from the relay to the receiver for Case B as $\hat{R}_1$, $\hat{R}_2$ and $\hat{R}_3$. Similar as the analysis for Case A, we have

$$\hat{R}_1 = t_1 \log_2(1 + \gamma_1), \tag{5.9}$$

$$\hat{R}_2 = t_1 \log_2(1 + \gamma_2), \tag{5.10}$$

$$\hat{R}_3 = t_2 \log_2(1 + \hat{\gamma}_3 \frac{t_0}{t_2}), \tag{5.11}$$

where $\hat{\gamma}_3 = \frac{\eta P h_{0,2} h_3}{\sigma^2}$. The instantaneous transmission rate of the user for Case B, denoted by $\hat{R}$, is formulated as

$$\hat{R} = \min\{\hat{R}_1 + \hat{R}_3, \hat{R}_2\}. \tag{5.12}$$

5.3 Throughput Maximization

In this section, we study the throughput maximization problems by finding the optimal time allocation scheme for both Cases A and B.

5.3.1 Relay with an Embedded Energy Source

We formulate the optimization problem for Case A as follows.

$$\max_{\tau} \quad R$$

$$\text{s.t.} \quad \text{C1: } \tau_1 + \tau_2 \leq T, \tag{5.13}$$

$$\text{C2: } \tau_1, \tau_2 \geq 0,$$

where $\tau = [\tau_1, \tau_2]$. Constraint C1 indicates that the summation of the user's backscattering time and the relay's forwarding time cannot exceed the duration of a transmission block, and constraint C2 limits that the optimization variables are non-negative. Denote the optimal solution for Problem P1 as $\tau^* = [\tau_1^*, \tau_2^*]$. Before solving Problem P1, we have the following lemmas.

Lemma 5.1 $R_3(\tau_2)$ *is an increasing function with respect to τ_2, and $R_1(\tau_1)$ and $R_2(\tau_1)$ are both increasing functions with respect to τ_1.*

Lemma 5.2 *In the optimal condition for (5.13), we have*

$$C3: \quad \tau_1^* + \tau_2^* = T, \tag{5.14}$$

$$C4: \quad R_1(\tau_1^*) + R_3(\tau_2^*) \leq R_2(\tau_1^*). \tag{5.15}$$

Proof Denote the optimal solution for Problem P1 as $\{\hat{\tau}_1, \hat{\tau}_2\}$, which satisfies $\hat{\tau}_1 + \hat{\tau}_2 < T$. Then, by contradiction, we show that $\{\hat{\tau}_1, \hat{\tau}_2\}$ is not the optimal solution. We consider that there exists $\{\tau_1^*, \tau_2^*\}$ satisfying

$$\tau_1^* + \tau_2^* = T, \quad \tau_1^* > \hat{\tau}_1, \quad \text{and} \quad \tau_2^* = \hat{\tau}_2.$$

From Lemma 5.1, we derive that $R_1(\tau_1^*) + R_3(\tau_2^*) > R_1(\hat{\tau}_1) + R_3(\hat{\tau}_2)$ and $R_2(\tau_1^*) > R_2(\hat{\tau}_1)$. It indicates that $\{\hat{\tau}_1, \hat{\tau}_2\}$ is not the optimal solution.

We further prove $R_1(\tau_1) + R_3(\tau_2) \leq R_2(\tau_1)$ in the optimal condition by contradiction. Denote the optimal solution for Problem P1 as $\{\tilde{\tau}_1, \tilde{\tau}_2\}$, where $\tilde{\tau}_1 + \tilde{\tau}_2 = T$. If $R_1(\tilde{\tau}_1) + R_3(\tilde{\tau}_2) > R_2(\tilde{\tau}_1)$, we can increase $\tilde{\tau}_1$ to τ_1^* and reduce $\tilde{\tau}_2$ to τ_2^* to guarantee that $R_1(\tau_1^*) + R_3(\tau_2^*) = R_2(\tau_1^*) > R_2(\tilde{\tau}_1)$ according to Lemma 5.1, where $\tau_1^* + \tau_2^* = T$. This contradicts with the assumption that $\{\tilde{\tau}_1, \tilde{\tau}_2\}$ is the optimal solution.

Therefore, we prove Lemma 5.2.

According to Lemma 5.2, we have

$$\max_{\tau} \quad R_1(\tau_1) + R_3(\tau_2) \tag{5.16}$$

$$\text{s.t.} \quad \text{C2, C3, C4.}$$

To obtain the optimal solution for (5.16), we consider the following two subcases. First, we consider the condition that $R_1(\tau_1^*) + R_3(\tau_2^*) = R_2(\tau_1^*)$, and derive the optimal solution for (5.16) in the following theorem.

Theorem 5.1 *If $R_1(\tau_1^*) + R_3(\tau_2^*) = R_2(\tau_1^*)$, the optimal solution for (5.16) satisfies*

$$\tau_1^* = T - \tau_2^*, \tag{5.17}$$

where $\tau_2^* > 0$ is the unique solution of $\tau_2 \log_2(1 + \frac{\gamma_3}{\tau_2}) + \tau_2 \log_2(1 + \gamma_2) - \tau_2 \log_2(1 + \gamma_1) = T[\log_2(1 + \gamma_2) - \log_2(1 + \gamma_1)]$, which can be easily obtained by the bisection method.

Then, we consider the condition that $R_1(\tau_1^*) + R_3(\tau_2^*) < R_2(\tau_1^*)$. For this condition, we have Theorem 5.2.

Theorem 5.2 *If $R_1(\tau_1^*) + R_3(\tau_2^*) < R_2(\tau_1^*)$, the optimal solution for Problem P2 satisfies*

$$\tau_1^* = T - \tau_2^*, \tag{5.18}$$

where $\tau_2^* > 0$ is the unique solution of $\log_2(1 + \frac{\gamma_3}{\tau_2}) - \frac{\frac{\gamma_3}{\tau_2}}{\ln 2(1 + \frac{\gamma_3}{\tau_2})} = \log_2(1 + \gamma_1),$
which can also be obtained by the bisection method.

From Theorems 5.1 and 5.2, we can conclude that both τ_1^* and τ_2^* are non-negative. That is to say, for Case A, the relay is always involved to forward the user's information transmission.

5.3.2 Relay with an Embedded Energy Source

For Case B, the optimization problem is given as follows.

$$\max_{t} \quad \hat{R}$$

$$\text{s.t.} \quad \text{C5: } t_0 + t_1 + t_2 \leq T, \tag{5.19}$$

$$\text{C6: } t_0, t_1, t_2 \geq 0,$$

where $t = [t_0, t_1, t_2]$. Denote the optimal solution for (5.19) as $t^* = [t_0^*, t_1^*, t_2^*]$. Similar as Lemma 5.2, we also have the following conditions for (5.19)

$$\text{C7: } \quad t_0^* + t_1^* + t_2^* = T, \tag{5.20}$$

$$\text{C8: } \quad \hat{R}_1(t_1^*) + \hat{R}_3(t_0^*, t_2^*) \leq \hat{R}_2(t_1^*). \tag{5.21}$$

From (5.20) and (5.21), we have

$$\max_{t} \quad \hat{R}_1(t_1) + \hat{R}_3(t_0, t_2)$$

$$\text{s.t.} \quad \text{C6, C7, C8.} \tag{5.22}$$

It can be proved that (5.22) is a convex optimization problem, which can be solved by the interior-point method [9]. However, this method needs iterations to find the optimal solution. To avoid the high-complexity iterations, we exploit the special structure of Problem P4 to obtain the optimal solution, for which Problem P4 is further decomposed into two sub-problems. First, given t_1, we find the optimal relationship between t_0 and t_2 by solving Problem P5.

$$\max_{t_0, t_2} \quad \hat{R}_3(t_0, t_2)$$

$$\text{s.t.} \quad t_0 + t_2 = T - t_1,$$

$$\hat{R}_3(t_0, t_2) \leq \hat{R}_2(t_1) - \hat{R}_1(t_1), \tag{5.23}$$

$$t_0, t_2 \geq 0.$$

The Lagrangian of (5.22) is given by

$$
\mathscr{L}(t_0, t_2, \lambda_1, \lambda_2) = \hat{R}_3(t_0, t_2) - \lambda_1(t_0 + t_2 - T + t_1)
$$
$$
- \lambda_2(\hat{R}_3(t_0, t_2) - \hat{R}_2(t_1) + \hat{R}_1(t_1)), \tag{5.24}
$$

where λ_1 and λ_2 are the Lagrangian multipliers, and the corresponding KKT conditions are given by

$$
\frac{\partial L}{\partial t_0} = (1 - \lambda_2)\frac{\hat{\gamma}_3}{\ln 2(1 + \hat{\gamma}_3 \frac{t_0}{t_2})} - \lambda_1 = 0, \tag{5.25}
$$

$$
\frac{\partial L}{\partial t_2} = (1 - \lambda_2)\left[\log_2(1 + \hat{\gamma}_3 \frac{t_0}{t_2}) - \frac{\hat{\gamma}_3 \frac{t_0}{t_2}}{\ln 2(1 + \hat{\gamma}_3 \frac{t_0}{t_2})}\right] - \lambda_1 = 0. \tag{5.26}
$$

Combining with (5.25) and (5.26), we have the following equation

$$
f(z) = \hat{\gamma}_3, \tag{5.27}
$$

where $f(z) = z \ln z - z + 1$, $z = 1 + \hat{\gamma}_3 \frac{t_0}{t_2}$. From [2], we know that $f(z)$ is an increasing function with respect to z ($z > 1$), and there is a unique solution $z^* > 1$ satisfying (5.27). Hence, we derive t_0 and t_2 with given t_1, which are given by

$$
t_0 = \frac{(z^* - 1)(T - t_1)}{z^* - 1 + \hat{\gamma}_3},
$$
$$
t_2 = \frac{\hat{\gamma}_3(T - t_1)}{z^* - 1 + \hat{\gamma}_3}.
$$

With the above result, we further derive t_1 by solving the following problem.

$$
\max_{t_1} \quad t_1 \log_2(1 + \gamma_1) + \frac{\hat{\gamma}_3(T - t_1)}{z^* - 1 + \hat{\gamma}_3} \log_2(z^*)
$$

$$
\text{s.t.} \quad 0 \le t_1 \le T, \tag{5.28}
$$

$$
\frac{\hat{\gamma}_3(T - t_1)}{z^* - 1 + \hat{\gamma}_3} \log_2(z^*) \le at_1,
$$

where $a = \log_2(1 + \gamma_2) - \log_2(1 + \gamma_1)$.

By solving (5.28), we have the following theorem.

Theorem 5.3 *By solving (5.28), the optimal solution is given by*

$$
\text{If } \log_2(1 + \gamma_1) \ge \frac{\hat{\gamma}_3 \log_2(z^*)}{\hat{\gamma}_3 + z^* - 1}, \quad t_1^* = T, \tag{5.29}
$$

$$if \ \log_2(1 + \gamma_1) < \frac{\hat{\gamma}_3 \log_2(z^*)}{\hat{\gamma}_3 + z^* - 1}, \quad t_1^* = \frac{b}{a+b}T, \qquad (5.30)$$

where $b = \frac{\hat{\gamma}_3 \log_2(z^*)}{\hat{\gamma}_3 + z^* - 1}$. *If* $\log_2(1 + \gamma_1) \geq \frac{\hat{\gamma}_3 \log_2(z^*)}{\hat{\gamma}_3 + z^* - 1}$, *it indicates that the system throughput maximum is obtained without the aid of the relay; If* $\log_2(1 + \gamma_1) < \frac{\hat{\gamma}_3 \log_2(z^*)}{\hat{\gamma}_3 + z^* - 1}$, *the relay is employed to maximize the system throughput, and* $t_0^* = \frac{(z^* - 1)(T - t_1^*)}{z^* - 1 + \hat{\gamma}_3}$, $t_2^* = \frac{\hat{\gamma}_3(T - t_1^*)}{z^* - 1 + \hat{\gamma}_3}$.

5.4 Numerical Results and Discussions

In this section, numerical results are provided to evaluate the performance of RaBackCom systems. The simulation parameters are given as follows unless stated otherwise. We set the transmit power of the CE as $P = 30\,\mathrm{dBm}$, the relay's embedded energy as $100\,\mu\mathrm{J}$, the noise power as $\sigma^2 = -70\,\mathrm{dBm}$, the relay's energy harvesting efficiency as $\eta = 0.8$, the transmission block duration as $T = 1\,\mathrm{s}$. The channel power gains are modelled as $h_{0,i} = 10^{-3}\theta_{0,i}d_{0,i}^{-\alpha}$ and $h_i = 10^{-3}\theta_i d_i^{-\alpha}$, $i = 1, 2, 3$, where $\theta_{0,i}$ and θ_i characterize the channel short-term fading and are set as $\theta_{0,i} = \theta_i = 1$ since we only consider the distance-dependent attenuation, and the path-loss exponent is set as $\alpha = 3$. We further set $d_{0,1} = d_{0,2} = 5\,\mathrm{m}$, $d_1 = 2\,\mathrm{m}$, $d_2 = 1.6\,\mathrm{m}$, and $d_3 = 0.7\,\mathrm{m}$. The scheme that BackCom without relay assistance is served as a benchmark.

Figure 5.3 shows the system throughput versus the relay's embedded energy. It is obvious that the throughput of Case A is larger than that of the benchmark. As the embedded energy increases, the throughput increases slowly. It is because for Case A, the time is mostly used for the user's information backscattering. Figure 5.4 investigates the system throughput versus the distance between the relay and the receiver. As the distance increases, the throughput reduces due to the decrease of h_3. Other observations are similar as Fig. 5.3.

Figures 5.5 and 5.6 investigate the performance of Case B. Figure 5.5 depicts the effect of CW signal's power on the system throughput. We observe that the throughputs of both Case B and the benchmark are increasing functions with the transmit power. For Case B, even a fraction of time is used for the relay's energy harvesting, Case B still achieves a larger throughput. Figure 5.6 shows the throughput versus the distance between the relay and the receiver. When the distance is small, the performance of Case B is superior to that of the benchmark. When the distance exceeds a threshold ($d_3 = 1\,\mathrm{m}$), the relay is not required to forward the user's information transmission. This observation is coincident with our analysis in Theorem 5.3.

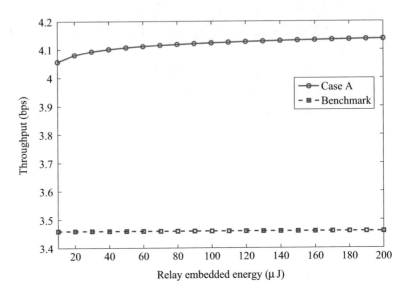

Fig. 5.3   Throughput vs. embedded energy

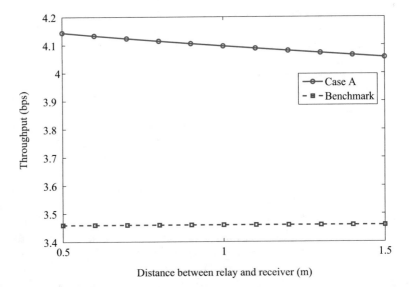

Fig. 5.4   Throughput vs. distance between relay and receiver for Case A

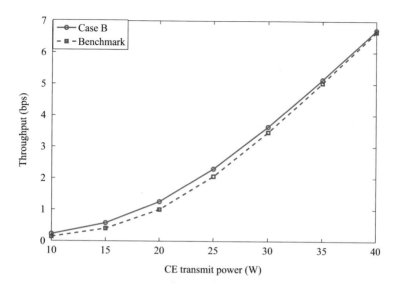

Fig. 5.5 Throughput vs. CE transmit power

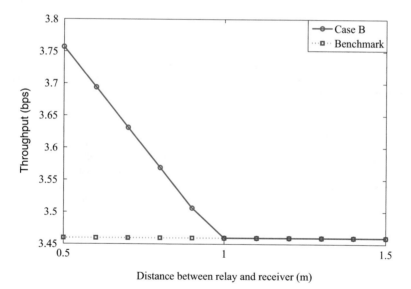

Fig. 5.6 Throughput vs. distance between relay and receiver for Case B

5.5 Conclusions

In this paper, we have considered a relay cooperation scheme in BCSs, where the relay is employed to forward the user's information to the receiver for the system throughput improvement. We have considered two cases that the relay is with/without an embedded energy source. If the relay has an energy source, it uses its own energy to forward the user's information. If not, the relay first harvests energy from the CW signal and then uses the harvested energy for information forwarding. For both cases, optimization problems for maximizing the system throughput have been formulated and closed-form solutions have been given, respectively.

References

1. Bin Lyu, Zhen Yang, Tianyi Xie, Guan Gui, and Fumiyuki Adachi, "Optimal time allocation in relay assisted backscatter communication systems," in *Proc. VTC-Spring*, Porto, Portugal, June 2018.
2. H. Ju and R. Zhang, "Throughput maximization in wireless powered communication networks," *IEEE Trans. Wireless Commun.*, vol. 13, no. 1, pp. 418–428, Jan. 2014.
3. H. Ju and R. Zhang, "Uer cooperation in wireless powered communication networks," in *Proc. GLOBECOM*, Austin, TX, USA, Dec. 2014, pp. 1430–1435.
4. C. Boyer and S. Roy, "Backscatter communication and RFID: Coding, energy, and MIMO analysis," *IEEE Trans. Commun.* vol. 62, no. 3, pp. 770–785, Mar. 2014.
5. H. Chen, Y. Li, J. L. Rebelatto, B. F. Uchäźĺ-Filho, and B. Vucetic, "Harvest-then-cooperate: Wireless-powered cooperative communications," *IEEE Trans. Signal Process.*, vol. 63, no. 7, pp. 1700–1711, Apr., 2015.
6. J. Qian, F. Gao, and G. Wang, "Signal detection of ambient backscatter system with differential modulation," in *Proc. IEEE ICASSP*, Shanghai, China, Mar. 2016, pp. 3831–3835.
7. X. Kang, Y. Liang, and J. Yang, "Riding on the primary: A new spectrum sharing paradigm for wireless-powered IoT devices," in *Proc. IEEE ICC*, Paris, France, May 2017, pp. 1–6.
8. Y. Liang and V. V. Veeravalli, "Gaussian orthogonal relay channels: Optimal resource allocation and capacity," *IEEE Trans. Inform. Theory*, vol. 51, no. 9, pp. 3284–3289, Sept. 2005.
9. S. Boyd and L. Vandenberghe, *Convex Optimization*. Cambridge University Press, 2004.

Chapter 6
Summary

In this chapter, we present a summary of main ideas of this book and discuss the future research directions about WPCN.

6.1   Summary of the Book

RF-based WPT technology can provide a stable energy supply to wireless devices. Considering RF signals carry both energy and information, the effective combination of WIT and WPT can solve the energy constraint problem of wireless devices, and promote the application of IoT. As one of the most important directions of WPC, WPCN has received a great deal of research interests from academia and industry. Recently, a lot of progress has been made in the research of WPCN, but there are still many problems remained to be solved. The following summarizes the main contributions of this book.

- An overview of WPCN is given. First, we overview the research background. Then, the state of art of WPCN is reviewed and discussed. Finally, we briefly discuss the basic working modes of WPCN, which includes the HTT mode and the BackCom mode.
- A practical NOMA-WPCN under SIC constraints is investigated. For the given system model, we propose a optimal time and energy allocation scheme to maximize the system throughput. From the derived results, we show that the distinctness among users' channel power gains can affect the system throughput under SIC constraints.
- Two hybrid schemes are proposed to exploit the advantages of the HTT and BackCom modes in WPCN. First, we study the hybrid user scheme, following which a WPCN has the users in the HTT mode and the users in the BackCom mode. Then, we design the hybrid mode scheme, where each user in WPCN

G. Gui, B. Lyu, *Optimization for Wireless Powered Communication Networks*,
SpringerBriefs in Electrical and Computer Engineering,
https://doi.org/10.1007/978-3-030-01021-8_6

can adopt the HTT mode or the BackCom mode. For each scheme, we design the optimal time allocation scheme to maximize the system throughput. For the hybrid user scheme, only the BackCom user with the largest backscatter rate can be scheduled. For the hybrid mode scheme, the optimal working mode permutation is analyzed.

- A hybrid mode scheme is proposed for CWPCN, where each CU can work in the HTT mode, the AB mode or the BB mode with sharing the spectrum with the primary communication pair. Under the studied system model, we design the optimal control policy, which includes the optimal time allocation and the optimal working mode permutation, to maximize the secondary communication system throughput.
- A relay cooperation scheme is proposed for BCS, where the relay forwards the received signals backscattered by the user to the IR. We consider two scenarios that the relay is with an embedded energy source and is without an embedded energy source. If the relay has the embedded energy source, it uses its own energy for information forwarding; otherwise, it first harvests energy and then uses the harvested energy to help the user's information transmission. For each scenario, the optimal time allocation is design to maximize the system throughput.

6.2   Future Research Directions

This section presents a range of future research directions for WPCN. As discussed above, extensive research efforts have been devoted to WCPN for IoT, however various challenging issues still remain open at the time of writing. In what follows, we present several interesting research topics to be explored in the future.

- *Wireless powered relay network.* Relay techniques are extensively used in wireless communication. However, relays may be deployed in some inaccessible places. Hence, the research about wireless powered relays is a hot topic. Moreover, the energy transmission efficiency in WPCN is typically low due to the path-loss. How to improve the energy transmission efficiency in wireless powered relay network is an important research directions for the future application of WPRN.
- *UVA enabled WPCN.* The recent works about WPCN typically consider that HAPs are static, because of which the coverage of WPT and WIT is limited. Unmanned aerial vehicles (UAVs) serve as mobile HAPs can significantly improve the coverage of WPT and WIT. For future work, we will design the trajectories of UAVs to maximize the system throughput.
- *Large-scale WPCN.* In the future IoT application, large-scale low-power sensors will be deployed throughout our lives. Stochastic geometry is a useful tool to model the large-scale WPCN model, for which we will further study the hybrid mode scheme to maximize the system throughput.

Printed in the United States
By Bookmasters